T0419600

LAWS AND LEGISLATION

THE DRINKING WATER SYSTEM IMPROVEMENT ACT

LAWS AND LEGISLATION

Additional books and e-books in this series can be found on Nova's website under the Series tab.

LAWS AND LEGISLATION

THE DRINKING WATER SYSTEM IMPROVEMENT ACT

NOAH L. HOGAN
EDITOR

Library of Congress Cataloging-in-Publication Data

ISBN: 978-1-53617-170-9

Published by Nova Science Publishers, Inc. † New York

CONTENTS

PREFACE

The United States uses 42 billion gallons of water a day—treated to meet Federal drinking water standards—to support a variety of needs. The Safe Drinking Water Act (SDWA) not only contains Federal authority for regulating contaminants in drinking water delivery systems, it also includes the Drinking Water State Revolving Fund (DWSRF) program. The DWSRF was created to provide financing for infrastructure improvements of drinking water systems. Chapter 1 looks at our Nation's drinking water infrastructure structure and examine questions as to what is necessary for the Federal Government to do in the way of planning, reinvestment, and technical support of these systems to meet future needs. Drinking Water System Improvement Act of 2017 amended the Safe Drinking Water Act to improve public water systems as discussed in chapter 2.

Chapter 1 - This is an edited, reformatted and augmented version of Hearing Before the Subcommittee on Environment of the Committee on Energy and Commerce, House of Representatives, One Hundred Fifteenth Congress, First Session, Serial No. 115–33, dated May 19, 2017.

Chapter 2 - This is an edited reformatted and augmented version of Report of the Committee on Energy and Commerce United States Senate, Report No. 115–380, dated November 1, 2017.

In: The Drinking Water System Improvement Act
Editor: Noah L. Hogan
ISBN: 978-1-53617-170-9

Chapter 1

DRINKING WATER SYSTEM IMPROVEMENT ACT AND RELATED ISSUES OF FUNDING, MANAGEMENT, AND COMPLIANCE ASSISTANCE UNDER THE SAFE DRINKING WATER ACT [*]

Subcommittee on Environment, Committee on Energy and Commerce

Friday, May 19, 2017
House of Representatives,
Subcommittee on Environment,
Committee on Energy and Commerce
Washington, DC.

The subcommittee met, pursuant to call, at 8:30 a.m., in room 2123, Rayburn House Office Building, Hon. John Shimkus, (chairman of the subcommittee) presiding.

Present: Representatives Shimkus, McKinley, Barton, Murphy, Harper, Johnson, Hudson, Walberg, Carter, Walden (ex officio, Tonko, Ruiz, Peters, Green, McNerney, Dingell, Matsui, and Pallone (ex officio).

Staff Present: Grace Appelbe, Legislative Clerk, Energy/Environment Subcommittees; Ray Baum, Staff Director; Mike Bloomquist, Deputy Staff Director; Jerry Couri, Chief Environmental Advisor; Jordan Davis, Director of Policy and External Affairs; Wyatt Ellertson, Research Associate, Energy/Environment Subcommittees; Blair Ellis, Digital

[*] This is an edited, reformatted and augmented version of Hearing Before the Subcommittee on Environment of the Committee on Energy and Commerce, House of Representatives, One Hundred Fifteenth Congress, First Session, Serial No. 115–33, dated May 19, 2017.

Coordinator/Press Secretary; Adam Fromm, Director of Outreach and Coalitions; Tom Hassenboehler, Chief Counsel, Energy/Environment Subcommittees; Zach Hunter, Director of Communications; A.T. Johnston, Senior Policy Advisor, Energy Subcommittee; Alex Miller, Video Production Aide and Press Assistant; Dan Schneider, Press Secretary; Sam Spector, Policy Coordinator, Oversight and Investigations Subcommittee; Hamlin Wade, Special Advisor, External Affairs; Jeff Carroll, Minority Staff Director; Jacqueline Cohen, Minority Chief Environment Counsel; David Cwiertny, Minority Energy/Environment Fellow; Rick Kessler, Minority Senior Advisor and Staff Director, Energy and Environment Subcommittees; Alexander Ratner, Minority Policy Analyst; Andrew Souvall, Minority Director of Communications, Outreach and Member Services; and C.J. Young, Minority Press Secretary.

OPENING STATEMENT OF HON. JOHN SHIMKUS, A REPRESENTATIVE IN CONGRESS FROM THE STATE OF ILLINOIS

Mr. Shimkus. I would like to call the hearing to order.

And I want to thank our witnesses for joining us today.

First of all, I know it is early. The one thing that is certain about us in Washington, D.C., is that there is uncertainty around us. So because of other meetings scheduled and planned, we asked for you to come early. And I do personally appreciate it. And it shows you the interest of our colleagues that they are here this early, so that is great.

No matter how many miles you travel—first of all, we have got folks from as far away as Alaska and as close as Pennsylvania here. No matter how many miles you have traveled to be with us, we are grateful for the time and financial sacrifice you are making to share your expertise with us today.

I also want to mention that even though they did not send someone to present oral testimony, I appreciate the Environmental Protection Agency providing us with a written statement to include in our hearing record. I ask for unanimous consent.

Without objection, so ordered.

[The information appears at the conclusion of the hearing.]

Mr. SHIMKUS. And I am also pleased to announce that the Agency has agreed to take written questions from members for our hearing record. This is highly unusual but an essential step to making this hearing record as complete as possible. And we obviously consider the Agency an important player whose technical experience and input is critical to the quality of our work.

I now recognize myself 5 minutes for giving an opening statement.

Today, our panel continues its look broadly at our Nation's drinking water infrastructure structure and examine questions as to what is necessary for the Federal

Government to do in the way of planning, reinvestment, and technical support of these systems to meet future needs.

The discussion draft which is subject to the hearing is meant to build on the testimony from our last hearing to help our subcommittee think more precisely about what items should be prioritized for legislation and how they should be addressed in the legislation.

Importantly, the discussion draft is not a finite universe of all the issues that the committee is open to considering. It is a true baseline for conversation and an invitation for feedback or refinements or suggested alternative approaches and an opportunity to make the case for including additional issues.

I know that some of us here today are curious why one provision or another is not added. I hope we can talk about those things today. I suspect we might be able to find agreement on some of those issues after we have had some time to find out each other's objectives and reflect on the best way to balance the needs of water, consumers, providers, and program implementers.

Let me take a minute to explain some items in the discussion draft, why they are there.

Based on oral testimony and written responses for the record, the water utility groups that testified at the last hearing talked about the importance of partnerships for addressing growth and compliance issues. The discussion draft proposes language to allow contractual arrangements or management of engineering services that will get a water system into compliance.

Under questioning, many of the witnesses mentioned the important role that asset management can play in addressing short- and long-term water system needs but that mandating this requirement would be challenging. The discussion draft has states consider how to encourage best practices in asset management and has the EPA update technical and other training materials on asset management.

We received testimony on the need to further aid disadvantaged communities. The discussion draft increases the amount a state can dedicate to disadvantaged communities to 35 percent of their annual capitalization grant and permit states to extend loan payments for these communities by another 10 years.

We received testimony on the need to increase funding for the Drinking Water State Revolving Loan Fund and the Public Water System Supervision grant, but not specific recommendations about what a realistic number is or whether commensurate budgetary cuts will offset these increases.

In response to this, the discussion draft creates a 5-year authorization for appropriations of both these programs but leaves them blank to allow a greater and more specific conversation to occur. This will not be easy. Some of these conversations will be very difficult, but we will have to have them in an open and honest manner, but that is not new. Anyone who has been around our subcommittee for a while knows we have a reputation for tackling challenging issues.

As I said earlier, we are at the beginning of this journey with a discussion draft as a baseline, and we are not close to the finish line as of yet.

With that, I yield back my remaining time. And now I yield to my friend from New York, the ranking member, Mr. Tonko.

[The prepared statement of Mr. Shimkus follows:]

Prepared Statement of Hon. John Shimkus

The Subcommittee will now come to order.

I want to thank our witnesses for joining us today, one from as far away as Alaska and the other as close as Pennsylvania. No matter how many miles you traveled to be with us, we are grateful for the time and financial sacrifice you are making to share your expertise with us today.

I also want to mention that, even though they did not have send someone to present oral testimony, I appreciate that the Environmental Protection Agency provided us written statement to be included in our hearing record, so ordered.

I am also pleased to announce that the Agency has agreed to taken written questions from Members for our hearing record. This is a highly unusual, but essential step to making this hearing record as complete as possible and we, obviously, consider the Agency an important player whose technical experience and input is critical to the quality of our work.

I now recognize myself for 5 minutes for giving an opening statement.

Today, our panel continues its look broadly at our nation's drinking water infrastructure and examine questions about what is necessary for the Federal government to do in the way of planning, reinvestment, or technical support of these systems to meet future needs.

The Discussion Draft, which is the subject of the hearing, is meant to build on the testimony from our last hearing and help our subcommittee think more precisely about what items should be prioritized for legislation and how they should be addressed in that legislation.

Importantly, the Discussion Draft is not a finite universe of the issues that the Committee is open to considering. It is a true baseline for conversation and an invitation for feedback on refinements or suggested alternative approaches and an opportunity to make the case for including additional issues.

I know that some of us here would are curious why one provision or another is not added, I hope we can talk about those things today. I suspect we might be able to find agreement on some of those issues after we have had some time to find out each other's objectives and reflect on the best way to balance the needs of water consumers, providers, and program implementers.

Let me take a minute to explain some items in the Discussion Draft and why they are.

Based on oral testimony and written responses for the record, the water utility groups that testified at the last hearing talked about the importance of partnerships for addressing growth and compliance issues. The Discussion Draft proposes language to allow contractual arrangements for management and engineering services that will get a water system into compliance.

Under questioning, many of the witnesses mentioned the import role that asset management can play in addressing short and long-term water system needs, but that mandating this requirement would be challenging. The Discussion Draft has States consider how to encourage best practices in asset management and has EPA update technical and other training materials on asset management.

We received testimony on the need to further aid disadvantaged communities. The Discussion Draft increases the amount a State can dedicate to disadvantaged communities to 35 percent of their annual capitalization grant and permits States to extend loan repayment for these communities by another 10 years.

We received testimony on the need to increase funding for the Drinking Water State Revolving Loan Fund and the Public Water System Supervision grants, but not specific recommendations about what a realistic number is — or whether commensurate budgetary cuts will offset these increases. In response to this, the Discussion Draft creates 5-year authorizations for appropriations of both of these programs, but leaves them blank to allow a greater and more specific conversation to occur.

This will not be easy — some of these conversations will be very difficult, but we will have to have them in an open an honest manner, but this is not new. Anyone who has been around our subcommittee for a while knows we have a reputation for tackling challenging issues.

As I said earlier, we are at the beginning of this journey – with the Discussion Draft a baseline — and we are not close to the finish line.

With that, I yield back my remaining time and now yield to my friend from New York, the Ranking Member of the Subcommittee, Mr. Tonko.

OPENING STATEMENT OF HON. PAUL TONKO, A REPRESENTATIVE IN CONGRESS FROM THE STATE OF NEW YORK

Mr. Tonko. Thank you, Chairman Shimkus. And thank you to our witnesses for being here on what is apparently a very busy morning in the House.

We can all agree that aging drinking water systems can hold back economic growth and threaten public health. These problems will only get worse if we continue the decade's long trend of neglect. I know we have limited time, so I will not restate all the details of our growing national need to invest in drinking water systems and update the Safe Water

Drinking Water Act. Suffice it to say, the need is immensely great. This subcommittee has been building a tremendous record that more than justifies the need for action.

Mr. Chair, I appreciate you holding this hearing and offering the discussion draft to bring attention to our hidden infrastructure, which has been out of sight and, regrettably, out of mind for far too long.

This draft responds to many of the issues that have been identified in previous hearings: the need to reauthorize the Drinking Water SRF and the Public Water System Supervision program; as well as the need to encourage asset management plans, greater source water protection, and support for disadvantaged communities.

With that said, I truly believe we can improve upon the draft before us today which will ensure strong bipartisan support moving forward. There are a number of democratic bills that have already been introduced that can help inform these efforts. The AQUA Act includes provisions on how to further assist disadvantaged communities and better incentivize asset management plans. It would also help fulfill a stated goal of this administration mandating Buy America requirements.

Mr. Pallone's SDWA amendments would enable EPA to promulgate much needed national standards. The bill also creates programs to reduce lead in schools among other important SDWA updates.

Mr. Peters has a bill to provide grants to assist systems with resiliency, source water protection, and security in the face of changing hydraulic conditions, such as droughts, sea level rise, and other emerging pressures on systems.

We do know the national need is growing: $384 billion over the next two decades to maintain current levels of services. We need to have the vision to acknowledge that this does not account for stresses, environmental and financial, that will continue to get worse if we simply do nothing.

Finally, the Drinking Water SRF has been a tremendous success. I am grateful that Chair Shimkus has undertaken the first funding reauthorization since its inception in 1996. But as we will hear today, the draft includes unspecified funding levels.

As a candidate, President Trump called for tripling funding for both SRF programs. The AQUA Act proposes levels that are in line with that—with what states handled following the Recovery Act. I think these are good targets to start negotiations.

We must recognize that local governments are struggling. Significant amounts of projects go unfunded each year, and the status quo of Federal support will simply not reduce the massive and growing levels of need. It is time for the Federal Government to step up and contribute its fair share.

Mr. Chair, I would end by asking for a commitment to sit down with our side, learn more about some of our proposals, and work together to make this a truly bipartisan effort that moves us forward. We had close cooperation on the brownfields reauthorization draft. I think we can get to a similar place on drinking water.

And with that, I yield back.

Mr. Shimkus. I thank the gentleman. The gentleman yields back.

The chair now recognizes the Chairman of the Full Committee, Mr. Walden, for 5 minutes.

OPENING STATEMENT OF HON. GREG WALDEN, A REPRESENTATIVE IN CONGRESS FROM THE STATE OF OREGON

Mr. Walden. Thank you, Mr. Chairman.

In March, our committee began a review of the financial needs of our entire Nation's drinking water infrastructure. We spoke about the need to think broadly about all things that can affect water affordability, reliability, and safety. Today, we take the next steps in our deliberative process by reviewing the discussion draft and related ideas from stakeholders to formulate policy on drinking water, state revolving loan funding, and Public Water System Supervision grants.

We will also examine efforts to improve asset management by utilities and other ways to lift paperwork burdens and improve systems delivery of safe drinking water.

Both sides of the aisle support making newer and larger investments in our Nation's infrastructure, and I agree that we need to help ensure these assets support the great quality of life Americans enjoy. However, in doing so, we must be careful to select wise investments and create diversified options that make sense for water systems, for states, and for consumers. It is important for us to tackle this job seriously for a couple of reasons.

As we learned at the last hearing, the country's drinking water delivery systems are facing the challenges of older age. We learned from the water utilities and other stakeholders the importance of partnerships for addressing growth and compliance issues.

The discussion draft proposes language to allow contractual arrangements for management and engineering services that will get a water system into compliance. We welcome feedback on that approach.

We also received testimony on the need to increase funding for the Drinking Water State Revolving Loan Fund and Public Water System Supervision grants, but not specific recommendations about what a realistic number is or whether budgetary cuts will offset these increases.

For the last couple of years, the appropriated levels have been consistent. The appropriations for the Drinking Water Revolving Loan Fund were last authorized in 2003. That is long enough. It is time to reassert this committee's proper role in authorizing our statutes and realign the focus of the EPA and other agencies back to their core missions, in this case, ensuring the provision of safe drinking water for our Nation's consumers.

We look forward to continuing the dialogue on this as our committee process continues.

I want to welcome all of you here today, our witnesses, who took time and traveled from far and wide to be with us to comment on this discussion draft, and that is what it is. Your input is important, and we would appreciate specific recommendations as you are able to give on these important issues.

And, again, thank you all for being here. We all care deeply about drinking water, safe drinking water, and helping our communities achieve that for all of our citizens in the country.

And with that, Mr. Chair, I yield back the balance of my time.

[The prepared statement of Mr. Walden follows:]

Prepared Statement of Hon. Greg Walden

In March, our committee began a review of the financial needs of our entire nation's drinking water infrastructure. We spoke about the need to think broadly about all the things that can affect water affordability, reliability, and safety.

Today we take the next steps in our deliberative process by reviewing a discussion draft and related ideas from stakeholders to formulate policy on drinking water state revolving loan funding and public water system supervision grants. We will also examine efforts to improve asset management by utilities, and other ways to lift paperwork burdens and improve systems' delivery of safe drinking water.

Both sides of the aisle support making newer and larger investments in our nation's infrastructure and I agree that we need to help ensure these assets support the great quality of life Americans enjoy in the future. However, in doing so, we must be careful to select wise investments and create diversified options that make sense for water systems, states, and consumers. It is important for us to tackle this job seriously for a couple reasons.

As we learned at the last hearing, the country's drinking water delivery systems are facing the challenges of older age. We learned from the water utilities and other stakeholders the importance of partnerships for addressing growth and compliance issues. The discussion draft proposes language to allow contractual arrangements for management and engineering services that will get a water system into compliance. We welcome feedback on that approach.

We also received testimony on the need to increase funding for the Drinking Water State Revolving Loan Fund and the Public Water System Supervision grants, but not specific recommendations about what a realistic number is – or whether budgetary cuts will offset these increases. For the last couple years, the appropriated levels have been consistent; the appropriations for the Drinking Water Revolving Loan Fund were last authorized in 2003. That is long enough. It is time to reassert this committee's proper role in authorizing our statutes, and realign the focus of the EPA and other agencies back to the core missions: in this case ensuring the provision of Safe Drinking Water for our nation's consumers. We look forward to continuing the dialogue on this as our process continues.

I want to welcome all our witnesses who took time and traveled from far and wide to be with us to comment on the discussion draft. Your input is important, and we would appreciate as specific of recommendations as you are able to give on these important issues.

Mr. SHIMKUS. The gentleman yields back the balance of his time.

The chair now recognizes the Ranking Member of the Full Committee, Mr. Pallone, for 5 minutes.

OPENING STATEMENT OF HON. FRANK PALLONE, JR., A REPRESENTATIVE IN CONGRESS FROM THE STATE OF NEW JERSEY

Mr. Pallone. Thank you, Mr. Chairman. Thank you.

The safety of our drinking water is an incredibly important topic which deserves more time than we have at today's hearing.

At our last drinking water hearing, we heard broad agreement from witnesses and members that we need to reauthorize the Drinking Water State Revolving Fund and increase the funding. My democratic colleagues and I have been saying this for years, so I am encouraged that Republicans on this subcommittee now seem to agree.

Unfortunately, this rushed hearing is not sufficient to address this issue. We have great ideas, but they are not reflected in the barebones discussion draft. We need a bipartisan effort to modernize the Safe Drinking Water Act, but in preparing this discussion draft, your staff didn't consult with us. We were eager to work with you, but we were told, without explanation, that such discussions could only happen after this hearing.

So before us today is a discussion draft that, in my opinion, fails to measure up to the severity of the problem. It simply does not meet the needs of public water systems and the communities they serve. The draft contains nothing to address the growing problems of lead in drinking water in homes and schools. It does nothing to improve the regulatory process and better protect public health from new and emerging pollutant classes, and it does nothing to improve transparency and restore consumer confidence in the safety of our tap water, and there is no commitment to increase funding.

So I am disappointed in the discussion draft, and I urge my colleagues to look at the real solutions in the bills that my democratic colleagues and I have introduced, and that is H.R. 1071, the AQUA Act of 2017, and H.R. 1068, the Safe Drinking Water Act Amendments of 2017.

I want to thank our witnesses for coming. I apologize that we don't have more time available, but I also want to express my frustration at the lack of a witness from the EPA. This subcommittee cannot produce meaningful legislation to reauthorize the state revolving fund and strengthen the Safe Drinking Water Act without their input. So it is clear we need to have another hearing.

Safe drinking water is simply too important, and I hope we can start to work together on a bipartisan bill to tackle these serious problems.

I yield back, Mr. Chairman.

Mr. Shimkus. The gentleman yields back his time.

All members having concluded their opening statements, the chair would like to remind members that pursuant to the Committee rules, all members' opening statements will be made part of the record.

I want to thank all of our witnesses for being here today and taking the time to testify before the subcommittee. Today's witnesses will have the opportunity to give opening statements, followed by a round of questions from members. Our witness panel for today's hearing are in front of us.

What I will do is recognize you individually for 5 minutes. Your full statements are submitted for the record. And as you can see, there is a lot of interest from our side. So if you get too far over the 5 minutes, I might start tapping the gavel to get you to wind up.

And before I take more time, let me just start by recognizing Mr. Martin Kropelnicki, President and CEO of the California Water Services Group, on behalf of the National Association of Water Companies. He testified here before. We are glad to have you back.

You are recognized for 5 minutes.

Statements

Martin A. Kropelnicki, President And Ceo, California Water Service Group, On Behalf Of The National Association Of Water Companies; Scott Potter, Director Of Nashville Metro Water Services, Nashville, Tn, On Behalf Of The American Municipal Water Association; Steve Fletcher, Manager, Washington County Water Company, Nashville, Il, On Behalf Of The National Rural Water Association; Lisa Daniels, Director, Bureau Of Safe Drinking Water, Pennsylvania Department Of Environmental Protection, On Behalf Of The Association Of State Drinking Water Administrators; Kurt Vause, Special Projects Director, Anchorage Water And Wastewater Utility, On Behalf Of The American Water Works Association; Lynn Thorp, National Campaigns Director, Clean Water Action; And James Proctor, Senior Vice President And General Counsel, Mcwane, Inc.

Statement of Martin A. Kropelnicki

Mr. Kropelnicki. Thank you, Mr. Chairman.

Good morning. I am Marty Kropelnicki, President and CEO of California Water Service Group, or Cal Water. We provide water and wastewater services to approximately

2 million people in the great State of California, Hawaii, New Mexico, and Washington, State of Washington. I am also the current President of National Association of Water Companies, which I am here representing today. NAWC's members have provided water and utility services for more than 200 years, and they serve approximately 25 percent of the U.S. population.

NAWC applauds you, Mr. Chairman, and this subcommittee for highlighting America's drinking water infrastructure needs and putting forward a discussion draft amendment to the Safe Drinking Water Act for utilities and regulators to review.

We are all working together toward the same outcome: safe, reliable, sustainable high-quality drinking water, which is critical to every person, every community, and every business in this country.

Suffice it to say that substantial portions of the utility sector face significant challenges. The Nation's drinking water infrastructure recently received a D by the American Society of Civil Engineers. The American Water Works Association projects that $1 trillion will be needed to invest infrastructure through 2035 to replace aging infrastructure to keep up with population growth.

More ominously, recent reports by the Natural Resources Defense Council showed that nearly one in four Americans get drinking water from untested and contaminated systems.

With great challenges come great opportunities, and that is what we are here to talk about today. The discussion draft put forward by the subcommittee is a good first step to addressing the crisis. Legislation along these lines would do much to build upon and advance the good work of many water suppliers that are already undertaken.

For example, NAWC estimates that our six largest members, of which Cal Water is one, will invest in nearly $2.7 billion this year alone in their water systems to ensure that they remain safe, reliable, and are sustainable for decades to come.

Federal funds alone will not fix this problem, especially given that many of the problems are the results of poor decisionmaking year after year after year and not necessarily the absence of funding.

Let me highlight for you several recommendations for Congress to consider. First, we must ensure that any Federal funds are used efficiently and effectively. NAWC and its members support the EPA's 10 attributes of effective utility management, which includes things such as financial viability and infrastructure stability.

Applicants for dollars of public funds should demonstrate that there are management assets that adequate repair, rehabilitation, and replacement are fully reflected in management decisions, including water rates that reflect the true and full cost of service.

Second, failing systems that are in seriously noncompliant situations with water quality standards must be held accountable. If a system is plagued with a history of serious noncompliance, it should be given an option to pursue a partnership that will lead to compliance or be compelled to consolidate system with an able owner or operator.

Finally, as Congress considers future funding for drinking water programs, NAWC recommends that the private water sector not only have equal access to Federal funding but also that steps be taken to further enable and incentivize private water investment and involvement in solving the Nation's infrastructure challenges.

Apart from the more obvious tax-based measures, these incentives should include providing a safe harbor or a shield that would allow companies like Cal Water or NAWC members to partner with undercompliant systems and give them that ramp-up time to be coming into compliance.

Quite simply, private water companies like Cal Water and NAWC members have the financial balance sheets, managerial and technical expertise to help ensure that all Americans have safe, reliable, and sustainable high-quality drinking water.

I sincerely appreciate the invitation to come back here today to testify. Along with my colleagues at NAWC, we look forward to continuing our work with you and this committee as we work on the Nation's infrastructure challenges.

Thank you. And I would be happy to respond to any questions, Mr. Chairman.

[The prepared statement of Mr. Kropelnicki follows:]

Testimony of Martin A. Kropelnicki

President and CEO, California Water Service Group

President, National Association of Water Companies

H.R._______, Drinking Water System Improvement Act and Related Issues of Funding, Management, and Compliance Assistance Under the Safe Drinking Water Act.

Presented on behalf of the National Association of Water Companies

House Energy and Commerce Committee

Subcommittee on Environment

May 19, 2017

Good morning, Chairman Shimkus, Ranking Member Tonko, and Members of the Subcommittee. Thank you for the opportunity to discuss efforts to improve the federal Safe Drinking Water Act. We appreciate the opportunity to comment on the Committee's discussion draft legislation.

I am Marty Kropelnicki, President and CEO of California Water Service Group (Cal Water), the third largest publicly traded water and wastewater utility company in the United

States. I had the pleasure of appearing before this Subcommittee in March to discuss the challenges facing the water industry and the nation's drinking water infrastructure. I am pleased to be here today to continue exploring solutions to these challenges.

I am also thc President of the National Association of Water Companies (NAWC) – the association that represents the regulated private water utility service industry and professional water management companies. NAWC's core belief is that by embracing the powerful combination of public service and private enterprise, we can improve our nation's water infrastructure, and by doing so, ensure that all Americans and future generations have access to safe, reliable, and high-quality water utility service.

NAWC applauds you, Mr. Chairman, and this Subcommittee for highlighting America's drinking water infrastructure needs and for putting forward a discussion draft of amendments to the Safe Drinking Water Act for utilities and regulators to review. We are all working toward the same outcome - safe, reliable, and high-quality drinking water, which is critical to every person, community, and business in this country. NAWC's members are proud to provide these services to our customers.

NAWC members are located throughout the nation and range in size from large companies that own, operate, or partner with hundreds of systems in multiple states to individual utilities serving a few hundred customers. Through various innovative business models, NAWC's members serve more than 73 million Americans, nearly a quarter of our country's population.

Cal Water, for one, provides water and wastewater service to approximately two million people in California, Hawaii, New Mexico, and Washington. Every day, Cal Water treats and delivers more than 320 million gallons of drinking water to our customers. For us, there is nothing more important than protecting our customers' health and safety, and working each and every day to ensure they have safe, high-quality water each time they turn on the tap.

There are two key areas that I will focus on today as they relate to the discussion draft before us. First, there is a need to embrace and enact effective utility management practices and accountability for all water systems – whether these systems are public or private. Second, there is a need to address those drinking water systems that are consistently non-compliant with federal health and safety standards. Working on these two critical areas can help improve the drinking water systems across the country while also ensuring that limited federal dollars are spent efficiently and wisely.

Private Utility Role in Meeting the Nation's Drinking Water Needs

Let me start by providing some background on the private utility role in meeting today's drinking water needs. Private water systems have existed in the United States for more than 200 years. Today, the private water utility sector is highly regulated by state public utility commissions (PUCs), which set water rates; the U.S. Environmental Protection Agency, which sets federal drinking water quality standards; and various state agencies, which are also responsible for setting water quality standards and protecting public health. The private water utility sector focuses on long-term planning by making appropriate and necessary infrastructure investments in our nation's communities. As a result, private water companies have a proven track record of consistently meeting the drinking water needs of consumers in many areas of the country.

The private sector is already helping overcome water infrastructure challenges facing the country. Ensuring the high standard of quality that private water companies deliver requires extraordinary amounts of capital investment. NAWC estimates that its six largest members alone are collectively investing nearly $2.7 billion each year in their water systems – and these six companies provide service to about six percent of the U.S. population. In Cal Water's case, we are budgeting to invest about $1 billion in our water systems over the next five years.

It is significant that six of NAWC's members are collectively investing nearly $2.7 billion in their water systems when one considers that the current total federal appropriation for the Clean Water and Drinking Water State Revolving Fund (SRF) programs is approximately $2.3 billion annually. One of the factors that enable the private water sector to undertake such significant levels of investment is outstanding credit ratings. In fact, the corporate credit ratings of some of NAWC's members are amongst the highest in the U.S. For example, Cal Water's first mortgage bonds are currently rated AA-, and Cal Water has the highest credit rating of any utility in the U.S., as rated by Standard & Poor's.

In addition to helping to ensure our customers have safe, reliable, and high-quality water utility service, NAWC members provide significant economic benefits to the communities we serve. We pay federal and state income taxes, local property taxes, local pump taxes, and permit fees for projects, all of which provide much needed revenue to all levels of government in the country. We hire local employees, and provide them with good-paying jobs and competitive benefits. We procure local goods and services. And to help ensure our medium- and long-term financial stability, our employees' retirement benefits are fully funded as required by Generally Accepted Accounting Principles. All of these things contribute to the economic multiplier effect that benefits the regions and communities that we serve.

Perhaps most importantly, NAWC's members work diligently with our public health and economic regulators to ensure that we meet federal and state water quality and customer service standards every day. For example, a review of the EPA's enforcement database shows that there are more than 2,000 public water systems in the country that are deemed serious violators of the nation's drinking water standards. Yet, not a single one of those systems is owned and operated by one of NAWC's members. This fact confirms earlier research conducted by American Water Intelligence, which found that the "compliance record of major companies in the private water utility sector has remained nearly spotless."[1]

In summary, the private water utility sector stands able, ready, and willing to partner with local and state governments, as well as the federal government, to help meet the challenges our nation's water infrastructure will face in the coming years and decades. In addition to supplying necessary capital, private water companies can leverage decades of experience solving complex water challenges to help bring new water infrastructure projects online more quickly and efficiently.

Effective Utility Management and Accountability

Our water infrastructure systems are the backbone upon which communities are built. Water service is a critical part of the physical platform of the U.S. economy. Not a single business in any community can be established, let alone survive and thrive, without a sustainable water supply. Communities must have reliable, resilient, and sustainable water infrastructure systems to attract and retain industry, business, and qualified workers. Simply put, capital investment in water infrastructure means job creation across the country.

Unfortunately, aging and deteriorating water systems threaten economic vitality and public health, and communities nationwide are faced with massive fiscal challenges to replace critical water and wastewater infrastructure and effectively manage their systems. After all, water systems are one of the most expensive assets for a community to maintain, and many municipally-owned utilities simply cannot afford to properly maintain, let alone improve and modernize, their infrastructure. They have a limited revenue base, which must be used to meet all of the needs of the community, including everything from street maintenance to public safety, not just water and wastewater services. Oftentimes, these fiscal challenges exacerbate the fact that many municipally-owned suppliers are not subject to stringent oversight of their operations and have not implemented best management practices designed to ensure the safety and reliability of the service they provide their customers.

NAWC and its members support the Environmental Protection Agency's (EPA) ten attributes of effective utility management endorsed by all major water and wastewater associations, including the American Water Works Association (AWWA), National Association of Clean Water Agencies (NACWA), Water Environment Federation (WEF), Association of Metropolitan Water Agencies (AMWA), Association of Drinking Water Agencies (ASDWA), and the Association of Clean Water Administrators (ACWA). The

[1] American Water Intelligence, "Data Show IOUs a Cut Above in SDWA Compliance," October 2012, p. 10.

attributes include priorities such as financial viability, infrastructure stability, and operational resiliency, which reflect the basics of financial, technical and operational capacity of sustainable utility management.

Failing and noncompliant water systems not only create a growing financial burden, but they pose significant risks to public health and the environment. The fact that there are thousands of water systems across the country in significant noncompliance with the nation's drinking water standards is both unacceptable and unsustainable. If we are to change the status quo, we must offer more "carrots and sticks" in the regulatory toolbox.

As a good first step, and as a general rule, applicants for public dollars should demonstrate that they have fully accounted for the long-term costs of their projects, including any risks inherent in construction, operations, and/or maintenance, and have selected the delivery model that provides the best long-term value to the water supplier's customers. For a community to maintain and improve the condition of its infrastructure, and to ensure its long-term safety and reliability, water utilities should be expected, at a minimum, to manage their assets based on a process where adequate repair, rehabilitation, and replacement are fully reflected in management decisions and fully accounted for in water rates.

On this latter point, it is important to note one of the core differences between regulated private water utilities, like Cal Water, and some of their public counterparts. The water rates charged by regulated private water utilities are set by state public utilities commissions to ensure they reflect the actual cost of service, including the cost of capital, as well the costs of operating, maintaining, and upgrading their water systems. Regulated utilities do not rely on other sources of revenue that are not related to the water system, such as sales or property tax revenue. Nor can money customers pay to receive water service be diverted to other uses, which too often happens in some municipal systems. Not only does this approach send an efficient price signal to customers, but it also helps to ensure that the utility remains financially stable and is able to maximize the efficiency and service life of its water system infrastructure.

As well, we would be wise to assess impediments to effective utility management resulting from local procurement processes. Public procurement today tends to overvalue low initial costs and undervalue future obligations, rewarding bidders who can build cheaply, rather than those who offer the best value over a project's lifecycle. The end result is oftentimes higher operations and maintenance costs, and as repairs go unaddressed, water system infrastructure fails prematurely, resulting in expensive rebuilds and threats to public health. This is unacceptable and fiscally irresponsible.

PARTNERSHIPS AND CONSOLIDATION

We appreciate the Committee's inclusion in the discussion draft of language related to the need for asset management plans. This is an important step but we would ask the Committee to consider a more robust approach.

Drinking water systems must be expected to maintain their assets and operations in compliance with health-based laws. One option to help struggling systems that is currently under discussion is to encourage these systems to pursue partnerships *in lieu* of traditional enforcement.[2] Alternatively, when a return to compliance is unlikely to occur, the State should compel the transfer of water systems assets and/or operational control over the water system to a supplier with a proven track-record of effectively operating, maintaining, and upgrading its water systems. NAWC recognizes that traditional enforcement tools, such as administrative orders and civil penalties, are not always appropriate or practicable. However, if we are to address the nation's drinking water challenges, we must expect failing systems to do things differently and, in terms of compliance with water quality regulations, all water suppliers – public or private – need to be held to the same standard.

In this regard, NAWC has been working closely with other water groups to promote legislation that would encourage partnerships, ranging from peer-to-peer support and public-private partnerships (P3s) to transfer and consolidation. We simply cannot continue to expect failing systems to change unless good decision-making is incentivized, bad decision-making is discouraged by holding utilities accountable, and federal funds are targeted in a way to ensure they are being used efficiently and cost- effectively.

While NAWC and its members are mindful of the socioeconomic and financial complexities associated with our nation's growing water crisis, water suppliers must be held accountable when their water systems fail. We should expect communities to proactively seek assistance and support or they should get out of the business of water provision. Year after year there is talk of the growing water crisis, yet little is done to actually address it and the number of customers exposed to water that does not meet minimum water quality standards continues to grow.

What is truly needed to address these kinds of compliance issues is a willingness to explore innovative solutions such as partnerships and incentivized consolidation. While many communities continue to clamor for more federal funding, more funding is not going to solve this growing crisis. In many cases, water system failures – be they related to water quality, reliability, or both – are not solely due to the absence of funding, but rather are directly attributable to the failure of proper governance, poor decision-making, and lack of stringent oversight.

There are numerous opportunities for these kinds of partnerships to be employed across the country. For example, there are currently several thousand public water systems that

---

[2] This approach has been endorsed by the Business Roundtable in a report release this week titled "Back to Business: A Blueprint for Renewing America's Infrastructure," available at: http://bit.ly/2pWRcf9.

the EPA has deemed serious violators of federal drinking water standards. Many of these communities are simply unable to address these violations on their own, and they would benefit from a partnership with either the private sector or even a neighboring municipality.

Several states have already made progress in effectively utilizing partnership and consolidations. Kentucky, for example, has been a national leader in incentivizing the consolidation of public water systems. Over the last several decades, Kentucky has been able to consolidate more than 2,100 public water systems to less than 400 systems today. Similarly, in 2015, California enacted Senate Bill 88 that authorized the State Water Board to require systems that consistently failed to meet public health standards to consolidate with other systems through physical or managerial consolidation.

We recognize there are small and rural communities where few, if any, viable partnership options exist. It is in these cases where federal funding and technical assistance can be the most beneficial. Doing more to encourage and incentivize partnerships and consolidation where they are viable would allow Congress to reprioritize federal funding and technical assistance toward those systems and communities where partnerships and consolidation are not viable.

While these types of public-private partnerships are, in many cases, an efficient and cost-effective solution, there are numerous impediments to their expanded use, including the legal and financial liabilities of distressed systems. Such liabilities for past noncompliance, which can range from hundreds of thousands to millions of dollars, can be a "poison pill" to a prospective new operator or owner. To solve this problem, Congress should consider providing a more robust legal "safe harbor" to encourage more private sector participation, including investment. Without such liability relief, significant amounts of private capital and investment remain on the sideline.

Performance and Full Access

All water suppliers in the country – whether they are government- or privately owned – are public service providers and their customers are comprised of taxpayers who fund programs such as the State Revolving Fund (SRF) programs. Despite this, there has been a long-standing prohibition against private entities from receiving Clean Water SRF funding for treatment works and, although the EPA does not prohibit such access to the Drinking Water SRF, no fewer than 12 states have adopted such blanket prohibitions.[3] Congress should seek to correct this imbalance by making future SRF funding contingent on states giving all water suppliers equal opportunity to apply for these funds.

---

[3] Alabama, Arkansas, Colorado, Georgia, Kansas, Louisiana, Mississippi, Nebraska, North Carolina, Oklahoma, Tennessee, and Wyoming. *See*, May 3, 2017, Congressional Research Service, *Drinking Water State Revolving Fund: Program Overview and Issues*, available at https://fas.org/sgp/crs/misc/RS22037.pdf (last visited May 13, 2017).

Additionally, in 2003, the EPA established its Four Pillars of Sustainable Infrastructure, one of which was full-cost pricing. The principle was established based on a 2002 Government Accountability Office (GAO) report that found that 29 percent of drinking water utilities were not generating enough revenues, and 43 percent of those received some form of federal or state grant or loan.[4] Further, more than one in four utilities failed to have plans to manage existing capital assets, and more than half of the utilities with plans did not cover all of their assets or omitted key elements, such as an assessment of capital conditions. Things have not changed over the last 15 years. In fact, the situation has only gotten worse and the infrastructure funding gap continues to widen.

Toward this end, we believe it is time that those utilities that receive federal assistance be expected to develop and implement a financial plan that covers not only capitalization costs, but operation and maintenance, and rehabilitation and repair costs. We must expect performance in terms of meeting federal and state standards, protecting public health, and providing cost-effective services, not more of the status quo. Failing systems should no longer be subsidized without an expectation of financial and operational viability.

Quite simply, full-cost pricing of water utility service is the single most important element of any strategy to improve the nation's drinking water infrastructure and compliance with the country's water quality standards. Full-cost pricing helps to ensure the financial viability of water suppliers, which then enables the supplier to undertake needed maintenance of and upgrades to its facilities, both of which play a critical role in the supplier's ability to provide safe and high-quality water to its customers.

This transition to full-cost pricing should, however, be accompanied by adequate financial support to assist economically distressed communities and low-income households. In this regard, Congress may wish to consider providing relief directly to challenged and low-income customers. Currently, federal funds flow directly to water utilities, which enables them to charge lower rates to all of their customers, including those who are not facing any type of economic hardship. A more efficient approach may be to transfer funds directly to challenged and low-income customers, similar to the Low Income Home Energy Assistance Program for gas and electric customers.

Conclusion

Our current water infrastructure crisis has been in the making for several decades, and it may take several decades to change the direction and right the ship. Today's dwindling resources and increasing demand for safe, reliable, and high-quality water require a fundamentally different approach than what we have taken over the last several decades.

---

[4] *See*, GAO Report, April 11, 2002, *Drinking Water Infrastructure, Information on Estimated Needs and Financial Assistance*, available at http://www.gao.gov/assets/110/109253.pdf (last visited May 13, 2017).

The discussion draft in front of the Committee today is a good first step to addressing this crisis. As outlined in my testimony, we have specific suggestions related to effective utility management, partnerships, and the future of the State Revolving Funds. We look forward to continuing to work with the Committee as this legislation as it works its way through the Congressional process.

I sincerely appreciate your invitation to appear before the Subcommittee today and, along with my many colleagues in the National Association of Water Companies, look forward to continuing our work with you to ensure that all Americans benefit from innovations in financing which improve the water infrastructure so essential to their quality of life. Thank you, and I would be happy to respond to any questions you may have.

Mr. SHIMKUS. Thank you. The gentleman yields back his time.

The chair now recognizes Mr. Scott Potter, Director of the Nashville Metro Water Services in Nashville, Tennessee, on behalf of the Association of Metropolitan Water Agencies.

You are recognized for 5 minutes, sir. Thank you.

STATEMENT OF SCOTT POTTER

Mr. Potter. Good morning, sir.

Chairman Shimkus, Ranking Member Tonko, and members of the subcommittee, the Association of Metropolitan Water Agencies, or AMWA, appreciates the opportunity to offer our thoughts today on the Drinking Water System Improvement Act of 2017.

I am Scott Potter, Director of Metro Water Services in Nashville, Tennessee. We provide drinking water services to 190,000 households and 200,000 sewer accounts in Nashville and Davidson County in Tennessee. I also serve as President of AMWA's Board of Directors. AMWA is an organization representing the Nation's largest publicly owned drinking water utilities, which collectively serve over 130 million Americans with quality drinking water. Our members support reauthorization of the Drinking Water SRF, and we appreciate that the legislation before the subcommittee today would do so for the first time in the program's history.

My written testimony has been submitted for the record. It includes more detailed feedback on the various sections of the legislation, so I will use my time today to speak more generally about the bill and AMWA's priorities for reauthorization of the Drinking Water SRF.

Simply put, we believe that the Drinking Water SRF is a valuable program. It should remain a cornerstone of Federal efforts to promote cost-effective water infrastructure financing to help communities protect public health and meet the regulatory requirements of the Safe Drinking Water Act.

We are pleased the Drinking Water System Improvement Act preserves the existing framework of the Drinking Water SRF, while making several targeted modernizations to the program and the Safe Drinking Water Act as a whole.

For example, the bill will leverage the expertise of large water utilities by encouraging them to enter into agreements to help in-need water systems correct, identify water quality violations, and carry out necessary management and administrative functions.

The bill also recognizes the importance of asset management by directing states to describe steps they will take to promote the adoption of effective asset management principles, practices, and how they will assist local utilities in training their staff to implement asset management plans.

We support these measures, though AMWA also believes utilities that have completed qualifying asset management plans should be rewarded with a degree of additional preference when they apply for Drinking Water SRF assistance.

The idea is not to make asset management plans mandatory or to exclude systems without asset management plans from receiving funding, but instead to incentivize all public water systems that seek SRF dollars to think holistically about the full life-cycle costs of their infrastructure.

As this legislation continues to develop, AMWA would like to recommend several additional points for consideration. Perhaps most importantly, the final bill should reauthorize the Drinking Water SRF at a level that recognizes the immense nationwide water infrastructure need and does not inadvertently constrain Congress' ability to fund the Drinking Water SRF at an amount that appropriately responds to these needs.

For example, initial versions of the fiscal year 2017 EPA appropriations bill approved by the House and Senate committees last year would have provided more than $1 billion for the Drinking Water SRF. Given the Nation's infrastructure needs and the apparent willingness of appropriators to provide this level of investment in the program, this legislation should authorize the funding level comfortably in excess of this figure.

Earlier this year, AMWA and other water sector stakeholders endorsed a call to double Drinking Water SRF funding to roughly $1.8 billion. So a figure in this vicinity would serve as a reasonable starting point for the new authorization level.

AMWA also supports expanding the Safe Drinking Water Act's definition of a disadvantaged community eligible for additional assistance to include a portion of the utility service area. The statute currently requires all the utility service area to meet the state's affordability criteria, but this is difficult to achieve for large metropolitan water systems that typically serve diverse populations that both have areas of affluence and also areas with concentrations of people in need.

By allowing defined portions of a large utility service area to be classified as disadvantaged, more individual in-need neighborhoods served by America's large water providers would become eligible for the same type of benefits that are already available to many small cities and towns throughout the country.

Finally, we support codifying the ability of recipients to use Drinking Water SRF funds for projects to improve the security of a public water system.

In 2014, Congress explicitly allowed the use of clean water SRF funds for security improvement projects at publicly owned treatment works. So we believe it is appropriate to formally extend the same ability to public water systems.

In closing, AMWA believes this legislation is a good starting point for efforts to reauthorize the Drinking Water SRF. We look forward to continuing to work with members of the subcommittee on this legislation, and I will be happy to answer any questions the committee may have.

Thank you, sir.

[The prepared statement of Mr. Potter follows:]

Testimony of Scott Potter

Director, Nashville Metro Water Services

On Behalf of the Association of Metropolitan Water Agencies

Before the U.S. House of Representatives Energy and Commerce Committee Environment Subcommittee

Hearing on the Drinking Water System Improvement Act of 2017

May 19, 2017

SUMMARY OF THE TESTIMONY OF SCOTT POTTER

- The Drinking Water State Revolving Fund has provided nearly $32.5 billion in funding assistance to communities across the nation through 12,827 individual assistance agreements over the past twenty years, but Congress has never reauthorized the program. The time to do so is now.
- The Drinking Water System Improvement Act would make a number of targeted updates to the DWSRF program to ensure maximum efficiency and flexibility for community water systems.
- AMWA supports provisions in the legislation that would reauthorize DWSRF appropriations, encourage states to promote asset management planning, reduce

duplicative regulatory requirements, and facilitate cooperative partnerships to help public water systems maintain compliance with SDWA standards.

- AMWA encourages the subcommittee to explore additional revisions to the DWSRF program, such as expanding the definition of "disadvantaged community" in the statute to include portions of service areas, and to codify the ability of community water systems to use DWSRF funds for facility security enhancements.
- AMWA stands ready to work with the subcommittee and all members of Congress to advance legislation that renews the federal commitment to investing in our nation's drinking water infrastructure.

Chairman Shimkus, Ranking Member Tonko, and members of the subcommittee: the Association of Metropolitan Water Agencies (AMWA) appreciates the opportunity to offer our thoughts today on the Drinking Water System Improvement Act of 2017.

I am Scott Potter, Director of Nashville Metro Water Services in Nashville, Tennessee. Metro Water Services provides quality drinking water to more than 190,000 households and more than 200,000 sewer accounts in Nashville and Davidson County, Tennessee. Our two drinking water treatment plants have a combined capacity of 180 million gallons of water per day. The drinking water is conveyed by a distribution system consisting of more than 3,000 miles of water main, and our largest pipe is five feet in diameter.

I also serve as president of AMWA's Board of Directors, a position I have held since 2015. AMWA is an organization representing the nation's largest publicly owned drinking water utilities, which collectively serve more than 130 million Americans with quality drinking water. Our members support reauthorization of the Drinking Water State Revolving Fund, and we appreciate that the legislation before the subcommittee today would do so for the first time in the program's history.

My colleague Rudy Chow of the Baltimore City Department of Public Works testified on behalf of AMWA during the subcommittee's March hearing on reinvestment and rehabilitation of drinking water systems, so the scale of our nation's water infrastructure challenge is well documented. By now the subcommittee is well aware that EPA data shows that our country's drinking water infrastructure requires $384.2 billion worth of investment over the next two decades just to maintain current levels of service. Members of the subcommittee also know that AMWA and the National Association of Clean Water Agencies have projected that water and wastewater utilities could spend nearly $1 trillion over the coming 40 years as they adapt to changing hydrological conditions such as extreme drought, more frequent intense storms, and rising sea levels. These startling figures are some of the strongest arguments we have in favor of ongoing federal support for the nation's drinking water infrastructure.

For these reasons, AMWA is pleased to see the subcommittee consider this discussion draft of the Drinking Water System Improvement Act. Most importantly, the bill would formally reauthorize funding for the Drinking Water State Revolving Fund for the first time since the program's creation in 1996. Since that time the DWSRF has delivered nearly $32.5 billion in funding assistance to communities across the nation through 12,827 individual assistance agreements – but it has been more than a decade since the program's original congressional authorization expired. Given that Congress made several reforms to the Clean Water SRF in 2014, and also authorized the Water Infrastructure Finance and Innovation Act (WIFIA) pilot program, the time is right for Congress to renew its commitment to the Drinking Water SRF as well.

While AMWA supports reauthorization of the DWSRF, the organization does not believe that the program, or the Safe Drinking Water Act as a whole, is in need of a top-to-bottom overhaul. Both programs work well in their current forms, though both would benefit from a number of targeted updates to ensure maximum efficiency and flexibility for community water systems. We are pleased that the Drinking Water System Improvement Act begins that process.

The following are AMWA's comments on specific components of the draft legislation.

Section 2: Contractual Agreements

Section two would improve drinking water quality and public health by encouraging knowledge sharing and collaboration between utilities. The provision would build on the Safe Drinking Water Act's existing consolidation incentive to allow state regulators or EPA to temporarily suspend enforcement actions for specific violations at a water system when another utility submits a plan to enter into a contractual agreement to take over significant management or administrative functions of that system. This section will encourage AMWA members and other large water systems, which often have extensive operational and institutional knowledge at their disposal, to contract with nearby smaller systems to correct violations while enjoying the same temporary suspension of enforcement that the statute already allows in cases where one out-of- compliance utility is fully consolidated with or acquired by another water system. As is the case with SDWA's existing consolidation incentives, any such contractual agreement undertaken pursuant to this section must be approved by the state or EPA in order for the enforcement relief to apply.

Section 3: Asset Management

AMWA strongly supports efforts to encourage public water systems to complete asset management plans, which we define as "an integrated set of processes to minimize the life-cycle costs of infrastructure assets, at an acceptable level of risk, while continuously delivering established levels of service."

Section three recognizes the importance of asset management by directing states to include as part of their capacity development strategies a description of how the state is encouraging water systems to adopt best asset management practices, and assisting local utilities in training their staff to implement asset management plans. The section also requires EPA to periodically update handbooks and training materials made available to public water system operators to reflect the latest thinking on the best practices for asset management strategies in the water sector.

AMWA supports these provisions because they will encourage states and EPA to promote effective asset management as broadly as possible, but we also believe that more can be done to incentivize the adoption of asset management methods by individual utilities. As AMWA testified in March, the association supports amending the DWSRF program to give public water systems that have completed qualifying asset management plans a degree of additional preference when they apply for DWSRF assistance. The idea is not to exclude systems without asset management plans from receiving SRF funding, but instead to encourage all public water systems that seek SRF dollars to use asset management planning to think holistically about the life-cycle costs of their infrastructure.

Section 4: Authorization for Grants for State Programs

This section would reauthorize expired funding for EPA to make grants to states to carry out public water system supervision programs. As AMWA and other water sector organizations wrote in an April 25, 2017 letter to congressional appropriators, public water system supervision programs "ensure that water utilities have the information, technology, and capabilities to meet their mandated regulatory responsibilities." AMWA supports this reauthorization.

Section 5: State Revolving Loan Funds

This section includes several updates to the DWSRF that should make the program even more appealing to public water systems that seek funding. One change would allow states to provide loan subsidies of up to 35 percent to support projects in disadvantaged communities that meet the state's affordability criteria, up from the current statutory cap of 30 percent. This will provide useful additional assistance to communities in need, but unfortunately the impact of this change is limited by the statute's existing definition of a "disadvantaged community" as "the service area of a public water system" that meets the state's affordability criteria. The requirement that the entirety of a utility's service area must meet the affordability criteria is difficult to achieve for large metropolitan water systems, which typically serve diverse populations that have both areas of affluence and areas with concentrations of people in need. This diversity of the ability to pay of households throughout the whole community often prevents disadvantaged community assistance from reaching pockets of utility service areas that, if they were served by their own water system, would easily qualify as disadvantaged under their state's criteria. AMWA therefore supports amending the statute's definition of "disadvantaged community" to include both entire water system service areas as well as portions of service areas. With this change, more in-need neighborhoods served by America's largest water systems would become eligible for the same type of additional subsidization to support necessary drinking water infrastructure projects as is already available to many small cities and towns throughout the country.

Section five of the discussion draft would also ease DWSRF repayment terms, allowing principal and interest payments to begin 18 months after completion of the project (up from one year under current law), and extending the amortization term to up to 30 years after substantial completion of the project, up from the current limit of 20 years. Additionally the section would allow 40-year amortization periods for projects carried out in disadvantaged communities. AMWA supports these changes and the increased flexibility they will bring, though we again note that expanding the definition of "disadvantaged community" to include a portion of a utility service area would ensure that this new flexibility is accessible to the greatest number of low-income communities nationwide.

Section 6: Other Authorized Activities

This section would allow states to use set-aside DWSRF funds to update source water assessments that were previously mandated by the Safe Drinking Water Act. Given that one of the most effective ways to protect drinking water quality is to prevent contaminants from entering source waters in the first place, AMWA believes this provision is a valuable update to the existing statute.

Section 7: Authorization for Capitalization Grants to States for State Drinking Water Treatment Revolving Loan Funds

This section represents the first funding reauthorization in the history of the DWSRF, and AMWA strongly supports renewing this commitment to the program. However, as the specific authorization levels remain undefined in the discussion draft, AMWA urges the subcommittee to insert in the final legislation figures that may serve as a point of aspiration for a congress that has, in recent years, allowed DWSRF funding amounts to level off.

As I previously stated, the nationwide drinking water infrastructure investment needs have been well documented, by EPA and others. Most recently in March the Environmental Council of States released an inventory of all fifty states top "ready to go" water and wastewater infrastructure projects that could benefit from SRF loans. The document showed $14.2 billion worth of water and wastewater projects nationwide that could move forward today with an infusion of SRF dollars – a figure that is more than five times the total amount of Drinking Water and Clean Water SRF funding that was appropriated by Congress for the 2017 fiscal year.

Against this backdrop of well-documented need, any new five-year DWSRF reauthorization established through this legislation must not inadvertently constrain Congress' ability to fund the program at a level that appropriately responds to these needs. For example, even though the final FY17 omnibus appropriations bill left DWSRF funding level at $863 million, earlier in the budget process House and Senate appropriators each approved versions of FY17 EPA funding bills that would have provided more than $1 billion for the DWSRF this year. Given the nation's infrastructure needs and the apparent willingness of appropriators to provide this level of investment in the DWSRF, this subcommittee should authorize a funding level comfortably in excess of this figure.

The subcommittee should also avoid constraining future DWSRF appropriations by making sure that the annual authorization level does not fall below the highest regular annual funding level that Congress has actually appropriated to the program in recent history. This mark came during the 2010 fiscal year when the DWSRF received $1.387 billion, so the annual authorization amount should exceed this figure as well.

AMWA notes that as a candidate last fall, President Trump called for tripling funding for both SRF programs at EPA. While his initial FY18 budget blueprint falls short of this goal, AMWA and other water sector stakeholders have endorsed calls to double DWSRF funding to roughly $1.8 billion. An annual figure in this vicinity could serve as a reasonable starting point for a reauthorized DWSRF.

Section 8: Demonstration of Compliance with Federal Cross-Cutting Requirements

This section has the potential to make the DWSRF even more attractive as a water infrastructure funding mechanism by allowing EPA to waive requirements that a funding recipient achieve and document compliance with a certain cross-cutting federal laws if the recipient is able to demonstrate compliance with an equivalent state or local statute. For example, several states have their own environmental review laws that apply to water infrastructure projects. If EPA were to determine that a state's requirements are at least equivalent to the standards of the federal National Environmental Policy Act, then a public water system applying for DWSRF assistance could demonstrate its compliance with the state-level law rather than documenting its adherence to the federal statute. This has the potential to reduce the paperwork burden on DWSRF applicants and help projects move more expeditiously through the application process. AMWA is eager to explore the degree of cost and time savings that could be achieved as a result of this provision.

Conclusion

Again, AMWA supports many of the DWSRF reforms that are included in this legislation, and we appreciate that the bill wisely avoids amending the Safe Drinking Water Act to modify the contaminant regulatory process or to insert artificial standard- setting deadlines into the statute. Conversely, we would suggest other provisions for inclusion, such as codifying the ability of public water systems to use DWSRF funds for water facility security enhancements, thus putting the program on par with the Clean Water SRF, which was amended in 2014 to allow the use of funds for security improvements at treatment works. Above all, we recommend a robust funding authorization level that will allow DWSRF investments to grow unimpeded in the coming years.

AMWA believes the Drinking Water System Improvement Act is a strong bill that makes meaningful progress toward solidifying the Drinking Water State Revolving Fund for success in the coming years. AMWA looks forward to continuing to work with members of the subcommittee on this legislation.

Thank you again for the opportunity to testify, and I would be happy to answer any questions you may have.

Mr. SHIMKUS. The gentleman yields back his time. The chair thanks him.

And now I would like to recognize Mr. Steve Fletcher, Manager of the Washington County Water Company, Nashville, Illinois, in the great State of Illinois, and in the great district of the 15th Congressional District of Illinois, on behalf of—who represents that? I don't know—of the National Rural Water Association. You guys got me off my game.

You are recognized for 5 minutes.

STATEMENT OF STEVE FLETCHER

Mr. FLETCHER. Good morning, Chairman Shimkus, Ranking Member Tonko, and members of the subcommittee. I am Steve Fletcher from rural Illinois in Washington County.

Rural Illinois and New York and the rest of America thank you for this opportunity to testify on drinking water infrastructure. Thank you, Congressmen Shimkus and Tonko, for your visits to your local small communities in your districts to tour and help with specific community water issues. This is very much appreciated.

I also need to thank Congressmen Harper, Tonko, and the sub- committee for passing the Grassroots Rural and Small Community Technical Assistance Act into law in the last Congress.

I am representing all small and rural community water supplies today through my association with the Illinois and National Rural Water Association.

Our member communities have the responsibility of supplying the public with safe drinking water and sanitation every second of every day. Most all water supplies in the U.S. are small. Ninety-two percent of the country's 50,366 drinking water supplies serve communities with fewer than 10,000 persons. Illinois has 1,749 community water systems and 1,434 serve less than 10,000 people. New York has 2,343, and 2,195 of those serve communities with less than 10,000 people.

My water system is a not-for-profit rural water system started by a group of farmers in the late 1980s who organized and built the water system using funding from the Federal Government that allowed these mainly farm families to receive safe, piped drinking water for the first time. Without the financial help from the Federal Government, we could never have afforded to have safe public water or even a public water utility.

Before the development of the rural water systems, rural households, including mine, relied on cisterns and private wells that were contaminated with nitrate so we couldn't drink the water.

We are pleased to endorse the subcommittee's legislation of the Drinking Water System Improvement Act of 2017. Small and rural communities support the use of these existing Federal infrastructure initiatives like the SRFs as the primary delivery mechanisms for any new Federal water infrastructure initiative. These initiatives all have specific provisions targeting Federal water subsidies to community water projects based on environmental and economic need. If some type of needs-based targeting is not specifically included in any new water infrastructure legislation, the funding will bypass rural America and be absorbed by large metropolitan water projects.

This bill accomplishes this objective. We support the bill's extended maximum loan duration and increase in the amount of additional subsidization to disadvantaged communities. Commonly, low-income or disadvantaged communities do not have the ability to pay back the loan, even with very low interest rates, and require some portion of grant funding to make the project affordable to the rate payers.

I would like to make two more related policy points with my remaining time. First, there is a misconception among some stakeholders that SRFs are for small and rural communities. SRFs have no limitation on size or scope of a water project. According to the EPA, most SRF funding is allocated to large communities. Approximately 62 percent of Drinking Water SRF funding is awarded to large communities, including numerous SRF projects that cost over $50 or $100 million. SRFs work for all sized water systems, and we are grateful for your support of the programs.

My final point is regarding local governmental choice in decisions of consolidation and privatization. The decision for any local government to privatize or consolidate should be determined at the discretion of local citizens. There is nothing inherently more efficient or more economical in the operation of our private water utility versus the public governmental water utility.

Regarding consolidation, rural water associations and systems like mine have assisted in more communities consolidating their water supplies than any program or organization. Again, when communities believe consolidation will benefit them, they eagerly agree with these partnerships. I have numerous examples from my own community which partners with six neighboring water utilities in various forms. We do not think any new Federal regulatory policy at expense of local government control and choice for privatization or consolidation would be beneficial to local communities or their citizens.

Thanks, Mr. Shimkus, for being such a good friend in support of rural America and to give us this opportunity today. I am happy to answer any questions.

[The prepared statement of Mr. Fletcher follows:]

TESTIMONY OF STEVE FLETCHER

On Behalf of the Washington County Water Company, Illinois Rural Water Association National Rural Water Association

Before The United States House Representatives

Subcommittee on the Environment

May 19, 2017

Subject: ''Drinking Water System Improvement Act of 2017''

INTRODUCTION

Good Morning Chairman Shimkus and Ranking Member Tonko and members of the Subcommittee. Rural Illinois, New York and the rest of America thank you for this opportunity to testify on drinking water infrastructure. And I would especially like to thank you, Congressmen Shimkus and Tonko, for your visits to your local communities in your districts to tour and help with specific communities' water issues. This has been very much appreciated in those communities.

I am Steve Fletcher from a very rural part of Illinois in Washington County. I am representing all small and rural community water and wastewater supplies today through my association with both the Illinois and National Rural Water Associations. Our member communities have the very important public responsibility of complying with all applicable regulations and for supplying the public with safe drinking water and sanitation every second of every day. Most all water supplies in the U.S. are small; 92% of the country's 50,366 drinking water supplies serve communities with fewer than 10,000 persons, and 80% of the country's 16,255 wastewater supplies serve fewer than 10,000 persons.

I am the general manager of the Washington County Water Company which is a non-profit rural water district started by a group of farmers in the 1980s. These farmers organized and built the water district using funding from the federal government that allowed these mainly farm families to receive safe, piped drinking water for the first time. Without the financial help from the federal government, we could have never afforded to have safe public water or even a public water utility.

We are governed by elected, volunteer board members that live in our service area. Before the development of the rural water districts, rural households, including mine, relied on cisterns and private wells that were contaminated with nitrates so we couldn't drink the water. We also relied on steel tanks that would catch the rain water off the roof and run it though some rocks to filter out sediment – and some farms were using water from their ponds with only some rudimentary treatment. None of these were good or safe options.

Over the last four decades, our little water district has grown to serve 4700 users through four separate small municipalities that have decided to partner with us for various reason which I will explain in a bit. We expanded project-by-project by laying new lines when we could secure the funding. Every few years we would extend water lines another 50-100 miles, allowing for an additional 200-300 homes to get drinking water for the first time. It took us ten years to grow and extend enough to service the president of the water district.

APPROPRIATE PARTNERSHIPS

We also partner with our neighboring town of Egypt which decided to get out of managing their own small water utility and gave the management responsibilities to us. We assumed all its assets and debts three years ago and now operate and manage Egypt's drinking water system as a satellite and separate public water system under our governance.

I wanted to highlight our various forms of partnerships with our neighboring communities including outright ownership of the town of Egypt, to selling wholesale water to the villages of Okawville and Radom, to providing partial operations to the Village of Ashley, and to our partnership with the Village of DuBois where we provide the operations, maintenance and compliance testing to the Village while it retains full local governmental control.

I note these partnerships to make the point that regionalization and consolidation of small communities' water systems are occurring and there is no current legal or structural impediment for this to occur. We support the concept and encourage these partnerships when it makes local economic sense because growing economies of scale result in lower cost to the consumer than operating independent water utilities. In the 1990s, it became apparent to the villages of Ashly and Okawville that it would be more economical to purchase water from us than what it would cost to upgrade their treatment plants – so they chose to partner with us.

The key ingredient in any successful consolidation is local support for the consolidation – and local control of when and how they choose consolidation. Rural Water has led or assisted in more communities consolidating their water supplies than any program, policy or organization. Again, when communities believe consolidation will benefit them, they eagerly agree with these partnerships. However, if communities are coerced to consolidate, one can almost guarantee future controversy. We urge you to allow local governments the authority to choose when to merge, consolidate or enter into a partnership. If a community is out of compliance with the Safe Drinking Water Act, civil enforcement can drive a community to a compliance solution. However, they should be able to choose their preferred compliance solution whether it be new treatment, regionalization, technical assistance, governmental changes, etc. We would be very concerned if the federal government expanded its regulatory reach into this traditionally local governmental authority.

"DRINKING WATER SYSTEM IMPROVEMENT ACT OF 2017"

We appreciate the Subcommittee's efforts to make modifications to the Safe Drinking Water Act to assist local governments with drinking water infrastructure funding and other forms of assistance in your legislation, the "Drinking Water System Improvement Act of 2017." We are pleased to endorse the bill for the following reasons and make some comments if the Subcommittee makes any modifications to the bill:

1. First, small and rural communities support the use of these existing federal infrastructure initiatives as the primary delivery mechanisms for any new federal water infrastructure initiative. These initiatives all have specific authorizing provisions that recognize that most water utilities are small and have more difficulty affording public water service due lack of population density and lack of economies of scale and have some targeting or prioritization of federal water

subsidies based on need. The state revolving loans achieve this principled objective by requiring that federal subsidies be targeted to the communities most in need based on their economic challenges combined with the public health necessity of the project. If rural and small town America is not specifically targeted in the legislation that would authorize and fund new water infrastructure initiatives, the funding will bypass rural America and be absorbed by large metropolitan water developments. The "Drinking Water System Improvement Act of 2017" accomplishes this objective by including targeting to disadvantaged communities and small communities with minimum set-asides, and prioritization of projects with the greatest environmental and economic need.

2. Second, we support the extended maximum loan duration up to 40 years. This extension can make the difference in a community being able to afford a project by lowering the repayment amounts to a level where the community can afford to service the debt. This change also makes the Drinking Water SRF consistent with other maximum loan terms in federal programs.
3. Third, we support the increase to 35 percent of the amount of additional subsidization to include forgiveness of principal that can be used in disadvantaged communities. Commonly, low income or disadvantaged communities do not have the ability to pay back a loan, even with very low interest rates, and require some portion of grant or principal forgiveness funding to make a project affordable to the ratepayers.
4. Fourth, the "Drinking Water System Improvement Act of 2017" includes no additional regulatory burden or new unfunded mandates on small and rural communities. Enhancing drinking water quality in small communities is more of a resource issue than a regulatory problem. Most small community non-compliance with the Safe Drinking Water Act and Clean Water Act can be quickly remedied by on-site technical assistance and education. The current EPA regulatory structure is often misapplied to small and rural communities because every community wants to provide safe water and meet all drinking water standards. After all, local water supplies are operated and governed by people whose families drink the water every day and people who are locally elected.

When Congressman Tonko's "Assistance, Quality, And Affordability Act," or Aqua legislation, was first introduced in 2010, we testified in favor of that legislation. We think some of the positive targeting contained in the AQUA bill has been included in the "Drinking Water System Improvement Act of 2017" and we appreciate that and thank you, Representative Tonko, for your continued efforts to make sure federal water funding is targeted to communities most in need.

We urge you to consider two additional provisions to the legislation that we believe would make it more effective in reaching and assisting communities facing some of the most challenging water infrastructure situations. For the past few years, the Interior Appropriations Subcommittee has been mandating in the EPA appropriations bill that states must use 20 percent of their drinking water SRF grant for making grants to disadvantaged communities. Please consider codifying this policy in the Safe Drinking Water Act to make it permanent and please consider increasing the 20 percent to a higher level to ensure grants are available to make the most necessary water projects in the most economically disadvantaged communities possible. Also, please consider authorizing a technical assistance initiative dedicated to helping under-resourced communities with the application process. Many communities simply have difficulty completing the necessary paperwork and working through the engineering process to successfully obtain funding from the available federal funding sources. Authorizing a technical assistance provision that would fund one person with expertise in grant writing and project completion in each state would allow all communities access to this shared resource that no single community could afford to employ full-time. We think such a program would cost approximately $6.5 million and should be implemented with similar authority through the Grassroots Rural and Small Community Water Systems Assistance Act.

114th CONGRESS, 2d Session, H. R. 5538, AN ACT, Making appropriations for the Department of the interior, environment, and related agencies for the fiscal year ending September 30, 2017.

Title II – Environmental Protection Agency State and Tribal Assistance Grants

Provided further, ...20 percent of the funds made available under this title to each State for Drinking Water State Revolving Fund capitalization grants shall be used by the State to provide additional subsidy to eligible recipients in the form of forgiveness of principal, negative interest loans, or grants (or any combination of these), and shall be so used by the State only where such funds are provided as initial financing for an eligible recipient or to buy, refinance, or restructure the debt obligations of eligible recipients where such debt was incurred on or after the date of enactment of this Act...

EXAMPLE OF CHALLENGING WATER INFRASTRUCTURE SITUATION

The Village of Neponset in Bureau County, Illinois, only has a population of 473 persons. It was already carrying a lot of debt for its water utility infrastructure when is was mandated to upgrade its wastewater utility, install new treatment to comply with the federal drinking water standard for radium, and finance the refurbishing of their water tower (approximately $1.5 million). The community had to raise their rates by $15 a month to approximately a total of $100 monthly. Community leaders are concerned the high cost of water service will result in more empty homes. All of the main three federal water funding sources have been very helpful in assisting the community and we are hopeful this assistance will keep the community viable. In addition to refinancing their existing debt to a longer loan duration of 30 years, the drinking water SRF funds has provided two loans to Neponset (one each for water and sewer), USDA has provided an additional loan for their sewer upgrade, and they were also able to qualify for a grant from the Community

Development Block Grant program. This is a good example of all the various funding agencies working cooperatively to address a small community in dire need.

Unfortunately, we don't have the magic solution for how to adequately fund the SRFs, increase funding for national water infrastructure, or find feasible ideas for new funding streams other than the traditional federal discretionary appropriations process. However, we are grateful for this committee's continued advocacy for appropriations for the SRFs each year and continued attention to water infrastructure challenges. We also will be relying on this committee to ensure that any new national infrastructure initiative does not bypass rural and small town America as it progresses in Congress.

TECHNICAL ASSISTANCE

I want to especially thank Congressman Harper, Tonko, and the Subcommittee for passing the Grassroots Rural and Small Community Water Systems Assistance Act into law in the last Congress. Small and rural communities want to provide safe water and meet all drinking water standards – and on-site technical assistance gives them the shared technical resource to achieve it. Most small community non-compliance with the Safe Drinking Water Act and Clean Water Act can be quickly remedied by on-site technical assistance and education. However, the assistance must come from someone they trust (a peer) who is willing to travel directly to the community, has technical expertise to remedy that specific community's issue with their specific treatment and infrastructure, and be available on-site at any time (nights, weekends, middle of winter, etc.). We have not been able to have that legislation, Public Law 114-98, control all the technical assistance funding in the Environmental Protection Agency (EPA) appropriations bill which is preventing that technical assistance funding from reaching rural Illinois, New York, and other states. Any assistance you can provide to correct this issue with the EPA Appropriations Subcommittee is greatly appreciated. The reason why this authorization and the similar drinking water authorization need to be specifically cited in the appropriations bill is because they contain a critical mandate that the EPA must follow Congressional intent and give preference to the type of technical assistance that small communities find to be most beneficial. Again, we would be grateful for any help in getting this message to the EPA Appropriations Subcommittee.

SMALL AND RURAL COMMUNITY ISSUES

When thinking about national water infrastructure proposals, please remember that most water utilities are small and have more difficulty affording public water service due to lack of population density and corresponding lack of economies of scale. The small community paradox in federal water policy is that while we supply water to a minority of the country's population, small and rural communities often have more difficulty providing safe, affordable drinking water and sanitation due to these very limited economies of scale and lack of technical expertise. Also, while we have fewer resources, we are regulated in the exact same manner as a large community; we outnumber large communities by a magnitude of 10-fold, and federal compliance and water service is often a much higher cost per household. In 2017, there are rural communities in the country that still do not have access to safe drinking water or sanitation due to the lack of population density or lack of funding – some exist in my own county.

Small community water infrastructure projects are more difficult to fund because they are smaller in scale – meaning numerous, very complicated applications have to be completed and approved compared to one large project. This is compounded by the reality that small communities lack the administrative expertise to complete the necessary application process – and perhaps lack the political appeal of some large cities as well.

Because water infrastructure is often less affordable (i.e., a much greater cost per household) in rural America, a water infrastructure project poses a greater financial risk compared to a metropolitan project and, very importantly, requires some portion of a grant, not just a loan, to make the project feasible. The higher the percentage of grants required to make a project work results in less money repaid to the infrastructure funding agency and a correlating diminution of the corpus fund.

State Revolving Loan Funds (SRFs)

There is a current misconception among some stakeholders that the SRFs have a limitation on size or scope of a water project and don't leverage federal dollars. States can currently leverage a smaller amount of water funding to create a much larger available loan portfolio. Similarly, states can use their federal SRF grants to leverage larger loan portfolios. According to the EPA, State SRF programs can increase funds through different types of leveraging such as:

- Using fund assets as collateral to issue tax-exempt revenue bonds;
- Using funds from one SRF program to secure the other SRF program against default through cross-collateralization;
- Using funds from one SRF program to help cure a default in the other SRF program through a short-term cross-investment; and
- Increasing disbursements to incrementally fund multiple projects within a capital improvement plan.

A 2015, Government Accountability Office (GAO) report on the state revolving funds found: "EPA tracks the amount of additional loans that are made because of leveraged bonds. States' Clean Water SRF programs have issued approximately $31.8 billion in loans with leveraged bonds, and states' Drinking Water SRF programs have made approximately $5.3 billion in additional loans with leveraged bonds..." [Source: State Revolving Funds, August 2015 GAO- 15-567]

Regarding the misconception some stakeholders are advancing that the SRFs have a limitation on size or scope of a water project, there is no size or scope limitation for water projects under the state revolving funds. According to EPA, most SRF funding is allocated to large communities:

- Approximately 72 percent of clean water SRF funding is awarded to large communities
  (EPA Clean Water State Revolving Fund Annual Review).
- Approximately 62 percent of drinking water SRF funding is awarded to large communities
  (http://www.epa.gov/ogwdw/dwsrf/nims1/dwcsizeus.pdf).

A simple review of projects funded by the SRFs show numerous projects that cost over 50 million dollars. It appears that the SRFs are used in every large water project in the country. This assertion should be verified by the EPA. The state of New York lists multiple projects funded by the drinking water SRF that cost over one billion dollars.

Clean Water Financing Proposed Priority System (FY2016) New Jersey Department of Environmental Protection (*link*).	
CAMDEN CITY $58,648,000	MIDDLESEX COUNTY $111,313,000
CAMDEN COUNTY $50,664,000	PASSAIC VALLEY SC $132,505,000
MIDDLESEX COUNTY $363,247,000	PASSAIC VALLEY $63,223,000
JERSEY CITY MUA $47,046,000	BELLMAWR BOROUGH $66,350,000
BAYSHORE RSA $5,894,000	EDISON TOWNSHIP $55,475,000
PASSAIC VALLEY SC $134,646,000	CAMDEN RED AGENCY $172,309,000
PASSAIC VALLEY SC $58,205,000	KEARNY TOWN $107,557,000
PASSAIC VALLEY SC $60,117,000	PENNSAUKEN TWNP $55,431,000
BERGEN COUNTY UA $54,172,000	SAYREVILLE ERA $50,664,000
PASSAIC VALLEY SC $63,223,000	
State Revolving Fund for Water Pollution Control Federal Fiscal Year 2016 New York State Department of Environmental Conservation (*link*).	
GREENWOOD LAKE, VILLAGE OF $62,021,000	ONEIDA COUNTY PHASE 5B $117,000,000
SOUTHAMPTON, VILLAGE $30,552,000	ONEIDA COUNTY PHASE 6A $110,600,000
CHEEKTOWAGA, TOWN OF $50,000,000	SUFFOLK COUNTY SW SD #3 $88,572,000
NASSAU COUNTY BAY PARK $50,951,925	SUFFOLK COUNTY RT 25 $76,230,000
NASSAU COUNTY BAY PARK $524,750,000	UTICA, CITY OF $105,304,000
ONEIDA COUNTY PHASE 2B $59,500,000	

Projects for New York City (NYCMWFA)	
WARDS ISLAND BRONX $64,091,406	NEWTOWN CREEK STP UP $112,331,279
WARDS ISLAND STP REHAB $102,655,400	NEWTOWN CREEK STP UP $169,975,528
BOWERY BAY STP MOD $50,412,000	NEWTOWN CREEK STP UP $140,983,576
BOWERY BAY STP UP $204,301,784	NEWTOWN CREEK STP UP $42,212,389
TALLMAN ISLAND STP UP $280,322,476	NEWTOWN CREEK STP UP $361,199,252
JAMAICA STP IMP JA-179 $57,267,070	NEWTOWN CREEK STP UP $589,360,645
26TH WARD, BB, TI, WI, $93,802,596	PUMP STATIONS CSO [CSO $183,867,577
26TH WARD STP IMP $51,101,400	CONEY ISLAND CREEK CSO $69,107,016
26TH WARD STP IMP $100,595,678	CONEY ISLAND CREEK CSO $48,351,415
NEWTOWN CREEK STP UP $45,933,272	NYC-WATERSHED NPS 319 $116,225,648

Final Intended Use Plan Drinking Water State Revolving Fund October 1, 2015- September 30, 2016 (*link*). **NEW YORK CITY**
Croton Filtration Plant (Phase 11 of 16479), $1,200,000,000
3rd City tunnel and shafts, dist press, $470,000,000
Catskill& Delaware UV Disinfection, Treatment Plant $1,400,000,000
CALIFORNIA, FISCAL YEAR 2015-2016 Clean Water Revolving Fund Intended Use Plan (*link*).
Sacramento Regional County Sanitation District Echo Water Project $174,380,875
Sacramento Regional County Sanitation District Echo Water Project $65,426,778
South Coast Water District Tunnel Stabilization & Sewer Rehabilitation $102,560,000
Hi-Desert Water District Wastewater Treatment and Water Reclamation $142,349,314
City of Malibu Civic Center Wastewater Treatment & Recycling Facility $41,900,000
Santa Margarita Water District Trampas Canyon Recycled Water $47,450,000
City of North Valley Regional Recycled Water Program $96,617,856
Monterey Regional Water Pollution Control Agency Groundwater $82,000,000
Eastern Municipal Water District Recycled Water Supply Optimization $114,031,280
Los Angeles, Advanced Water Purification Facility $451,000,000
Sacramento Regional County Sanitation District Echo Water Project $59,408,652
Sacramento Regional County Sanitation District Echo Water Project $711,032,393
City of San Luis Obispo Water Resource Recovery Facility Expansion $68,000,000
Ventura County Waterworks District No. 1 $50,000,000
San Jose, City of Digester and Thickener Facilities $86,350,000
Water Replenishment District of Southern California Groundwater $80,000,000
Upper San Gabriel Valley Municipal Water District Indirect Reuse $65,000,000
Los Angeles, City of Hyperion Treatment Plant Membrane $460,000,000
Palmdale Water District Palmdale Regional Groundwater Recharge $130,000,000
Sacramento Regional County Sanitation District Echo Water Project $484,585,422

Privatization

NRWA has not opposed water supply privatization in principle. However, corporate water (profit generating companies or companies paying profits to shareholders/investors) should not be eligible for federal taxpayer subsidies. Private companies argue that they have to comply with the same regulations. However, the distinction in mission between public and private is the core principle that should be considered. Public water utilities were and are created to provide for public welfare (the reason why public water continues to expand to underserved and nonprofitable populations). Any federal subsidy that is provided to a corporate water utility should be separated from subsidizing that company's profits.

Regarding EPA's suggestion that public-private partnerships may be a solution for small and rural water utility "challenges," we urge EPA to limit its policy and initiatives to compliance rather than promote water utility privatization. EPA should leave any decisions regarding privatization to the local citizens' discretion. The decision for any local governmental to privatize, including incremental privatization, should be determined at the discretion of local citizens. There is nothing inherently more efficient or more economical in the operation of a private water utility versus a public-governmental water utility. As the Government Accountability Office concluded in 2008, "There is no 'free' money in public-private partnerships." This observation is self-evident, along with the observation that private water utilities are inherently no more efficient that public water utilities. While we believe that maximizing profit is a noble virtue, we do not think that federal policy and initiatives should promote privatization of water utilities.

Regarding private or commercial funding as a source for investment in the country's water infrastructure, please know that there is currently no limitation on private or commercial investments in water utility infrastructure projects. Many water utilities currently rely on commercial or private investors (i.e., a local bank) for certain projects. However, many water infrastructure water projects would become unaffordable, like the communities cited earlier in my testimony, if they were to rely solely on commercial or private financing. This means that the ratepayers would not be able to afford their water bills if the total cost of the project were financed by the ratepayers. This dynamic is especially acute in low-income communities with expensive water utility infrastructure needs.

Congress has determined that there is a federal interest in subsidizing some of these water infrastructure projects based on need – the community's lack of ability to afford the project combined with the public health or environmental urgency of the project. Congress appropriates finite water funding subsidies and communities compete based on need for these limited federal subsidies.

Under the Clean Water Act and the Safe Drinking Water Act, the state revolving funds' (SRFs) application processes require the prioritization of funding awards based on a meritorious needs-based evaluation conducted by the states. Under the U.S. Department of Agriculture's (USDA) water infrastructure funding program, communities must demonstrate they don't have the ability to obtain commercial credit (the "credit elsewhere" test) and then they are only subsidized by the amount to make the project affordable to that specific community based on a ratio of water rates and local median household income. There are never enough federal subsidies to fund every project.

We have concerns with proposals to extend new subsidies or tax preferences to the private investment sector to support a new national infrastructure initiative:

- For private or commercial funding instruments to be able make projects more affordable by lowering interest rates, the federal government would have to offer some type of subsidy or tax-break to the private sector. This will have a cost to the federal government in decreased tax revenue or direct appropriations. If this cost is used to support the private sector, it will result in a transfer or circumvention of public (taxpayer) subsidies from the public (local governments under the SRFs, USDA, etc.) to the commercial or corporate sector. We believe that federal water project subsidies should be used for the public/governmental sector water infrastructure projects determined to be a federal priority worthy of public subsidy.

- Private infrastructure financing does not require the prioritization of projects based on need (economical and environmental) like the current government water programs. It is in the interest of the private financing sector to fund the projects that would have the highest return on investments. Therefore, if additional federal subsidies were used to subsidize the private sector, if would have the effect of redirecting federal subsidies from the projects with the greatest need (economical, public health and environmental) to the projects with least need.

Again, there is currently no limitation of commercial or private investment in water infrastructure; our concern is limited to providing a new subsidy to the private or commercial sector that could remain in a public sector dedicated to accomplishing federally identified priorities.

Mr. Shimkus. The gentleman yields back his time, and the chair thanks him.

And now I would like to turn to Ms. Lisa Daniels, director of the Bureau of Safe Drinking Water at the Pennsylvania Department of Environmental Protection, on behalf of the Association of State Drinking Water Administrators.

You are recognized for 5 minutes.

STATEMENT OF LISA DANIELS

Ms. DANIELS. Good morning, Chairman Shimkus, Ranking Member Tonko, and members of the subcommittee. Thank you for the opportunity to be here to discuss the status of our Nation's state drinking water programs.

I am also President-Elect for ASDWA, so I am very glad to be here to represent the organization.

Our members are on the front lines every day ensuring safe drinking water and protecting public health. Vibrant and sustainable communities, their citizens, and businesses, all depend on a safe and adequate supply of drinking water.

States oversee more than 152,000 public water systems and interact with them through a broad range of activities that are funded through two Federal funding sources. Of course, there is the Drinking Water State Revolving Loan Fund, but there is also the Public Water System Supervision program.

The vast majority of community water systems are in compliance with health-based standards. That is the good news. But what about those systems that struggle?

The Drinking Water SRF can provide solutions for struggling systems. At only 20 years old, it really is a remarkable success story. It has allowed states to fund projects to upgrade treatment plants, rehabilitate distribution systems, address our aging infrastructure, and it has been quite successful. In fact, states have been able to leverage Federal funding to fund more than 13,000 projects through the SRF.

A major component of the 1996 amendments was new statutory language that allow states to undertake what we call proactive measures. Funded through the set-asides, proactive measures such as operator training, technical assistance, and source water protection offer support for water systems as they strive to enhance their performance.

Water systems are encouraged to consider a range of options, including partnerships, which could be as simple as sharing a back-hoe or as complex as merging with a neighboring system. And the set-aside funds are available to support many of these activities.

I would like to share an example from my home state. The Stockton Water System was a very small 43-home community that was operating as an untreated, unfiltered, and unpermitted surface water system. We discovered this system in 2014 because of customer complaints.

The water was found to contain *E. Coli*, *giardia*, and *salmonella*. Traditional strategies and enforcement weren't working in this community. They really needed a different kind of assistance. We employed several capability enhancement programs in Stockton, including Capability Enhancement program, which provided the initial assessment and also provided onsite technical assistance to really help folks understand the challenges with this community. We also employed the professional engineering services program, which was able to conduct feasibility studies and design work to find the best solutions.

These initiatives came together with PENNVEST, which is our SRF funding agency, to identify a willing partner. And we found that in the nearby Hazelton City Authority system. They agreed to work with Stockton, make the Drinking Water SRF application, extend water service, replace Stockton's existing distribution system, while keeping water rates at an affordable $35 per month. The total project cost was $2.2 million, which was underwritten by PENNVEST and, today, Stockton now has a safe and reliable supply of drinking water.

Solutions such as this would absolutely not be possible without the Drinking Water SRF and the set-asides. Drinking water systems and the communities they serve are the direct beneficiaries of the work accomplished through these programs.

State drinking water programs have often been expected to do more with less, and we have always responded with commitment and integrity, but we are currently stretched to the breaking point. Insufficient Federal funds increase the likelihood of contamination incidents, and we do not want to see another Charleston, West Virginia; Toledo, Ohio; or Flint, Michigan.

To sustain public health protection, states need congressional support. For the past 4 years, the PWSS program has flat funded, and the Drinking Water SRF funding has decreased. These essential programs come with well-documented needs, and they must be fully supported.

ASDWA recommends the PWSS program be funded at $200 million, and we also recommend the Drinking Water SRF be funded at $1.2 billion to allow us to continue to do this great work.

In summary, the 1996 amendments offered the community a promise of enhanced public health protection through a framework of both traditional and proactive collaboration between state drinking water programs and the water systems that they oversee. Maintaining funding for the Drinking Water SRF, the set-asides, and the PWSS program is critical.

State drinking water programs are committed to fulfilling the promise of the 1996 amendments. Thank you.

[The prepared statement of Ms. Daniels follows:]

"Drinking Water Systems Improvement Act"
Subcommittee on Environment and the Economy
House Energy and Commerce Committee
Friday, May 19, 2017
By Lisa Daniels
Director, Bureau of Safe Drinking Water Pennsylvania Department of Environmental Protection And President-Elect Association of State Drinking Water Administrators (ASDWA)

Good Morning Chairman Shimkus, Ranking Member Tonko, and Members of the Subcommittee. Thank you for this opportunity to talk about our Nation's drinking water systems and how state drinking water programs support them. My name is Lisa Daniels and I am the Director of the Bureau of Safe Drinking Water at the Pennsylvania Department of Environmental Protection. I am also the President-Elect of the Association of State Drinking Water Administrators (ASDWA), whose 57 members include the 50 state drinking water programs, five territorial programs, the District of Columbia and the Navajo Nation. Our members and their staff are on the front lines every day, ensuring safe drinking water and protecting public health. Their technical assistance and support, as well as oversight of the drinking water systems, are critical to providing safe drinking water and protecting public health.

Today, I'd like to talk with you about how the 1996 Safe Drinking Water Act (SDWA) Amendments have increased compliance for public water systems, the challenges that remain, and the tools and the resources that can be used for continuous improvement of the nation's water supply. My remarks will focus on the proactive elements of the 1996 Amendments and how these programs have built a framework of cooperation between water systems and state primacy agencies as well as enhanced performance by public water systems through training, education, outreach, and other support mechanisms.

OVERVIEW

For each of the 50 state drinking water programs, territorial programs, and the drinking water program of the Navajo Nation, our principal and enduring goal is public health protection. Vibrant and sustainable communities, their citizens, workforce, and businesses all depend on a safe, reliable, and adequate supply of drinking water. Economies only grow and sustain themselves when they have safe and reliable water supplies. More than 90% of the American population receives water used for bathing, cooking, fire protection, and drinking from a public water system – *overseen by state drinking water program personnel.* Public water systems rely on state drinking water programs to ensure they meet all applicable Federal requirements and the water is safe to drink.

THE 1996 SAFE DRINKING WATER ACT (SDWA) AMENDMENTS

To meet the requirements of the Safe Drinking Water Act (SDWA), states have accepted primary enforcement responsibility for oversight of regulatory compliance and technical assistance efforts for more than 152,000 public water systems to ensure that potential health-based violations do not occur or are remedied in a timely manner. To achieve this public health protection goal, states interact with the public water systems through a broad range of activities, including:

- Adopting Federal regulations or developing their own state-level regulations that are at least as stringent as the Federal regulations;
- Providing technical assistance and training for water systems on regulations, treatment, and technical, managerial, and financial issues;
- Reviewing plans and specifications for modifications to existing water systems and water infrastructure improvement projects;
- Inspecting water systems, including a review of all components from source to distribution and an audit of systems' record-keeping;
- Managing operator certification programs to ensure that treatment plant operators and distribution system personnel are appropriately certified and trained;

- Managing laboratory certification programs to ensure that the compliance monitoring analytical results are of the appropriate quality;
- Managing source water protection and capacity development programs;
- Managing water system security and preparedness programs;
- Reviewing applications and closing loans for the Drinking Water State Revolving Loan Fund (DWSRF);
- Managing compliance data and reporting results to EPA's Safe Drinking Water Information System (SDWIS);
- In some states, collecting compliance samples and conducting laboratory analysis; and
- When necessary to ensure compliance with public health standards, taking enforcement actions.

States accomplish this range of activities through two principal Federal funding sources – the Drinking Water State Revolving Loan Fund (DWSRF) and the Public Water System Supervision (PWSS) grant program. Taken together, these two Federal funding programs provide the means for states to work with their water systems to ensure that public health is protected. More than 90 contaminants are currently regulated under the SDWA and the vast majority of community water systems are in compliance with the health based-standards. In fact, in the years since the 1996 SDWA Amendments were enacted, the national compliance percentage with health-based standards has increased from 85% to 93% (by 2013, the most recent date for which data is available). But what about those systems that struggle?

THE DRINKING WATER STATE REVOLVING FUND (DWSRF)

We are all aware that one of the greatest challenges facing the drinking water community today is aging infrastructure. The DWSRF, although only 20 years old, is a remarkable success story. It has allowed states to award project dollars to utilities to help them upgrade their treatment plants, rehabilitate their distribution systems, install more protective technologies, and generally improve their aging infrastructure. Since its inception, the DWSRF has touched more than 852 million Americans through projects that enhance drinking water capabilities at water utilities. In the core DWSRF program, approximately $18.2 billion in cumulative Federal capitalization grants have been leveraged by states into over $32.5 billion in infrastructure loans to small and large communities across the country. 25.5% of the cumulative DWSRF assistance has been provided to disadvantaged communities. Such investments pay tremendous dividends – both in supporting our economy and in protecting our citizens' health. States have very effectively and efficiently leveraged Federal dollars with state contributions to provide assistance to more than 13,000 projects, improving health protection for millions of Americans. And, as described in the section below, DWSRF set-asides are an essential source of funding for states' core public health protection programs and work in tandem with infrastructure loans to support water system needs.

Drinking Water Set-Asides

A major component of the 1996 SDWA Amendments was new statutory language that allowed states, for the first time, to provide financial support for proactive measures in their work with drinking water systems to both support and meet their needs. These were all funded through what we call "set-asides" under the new Drinking Water State Revolving Loan Fund, the DWSRF. Specific percentages of a state's DWSRF can be set aside for programs relating to support for operator certification; enhancing a system's technical, financial, and managerial capabilities to attain and then sustain compliance with all applicable Federal requirements; and source water and wellhead protection initiatives; as well as training and technical assistance across all programmatic elements of implementing the SDWA. These proactive initiatives support water systems as they strive to enhance their performance and better protect public health. They also provide the financial wherewithal often not otherwise available for systems in need to meet their public health protection responsibilities. These programs offer the opportunity for states and systems to work together. Here's how they work...

Through the set-asides, states can provide training for operators. Programs can be designed to address specific needs from the simple, such as basic math, to the complex, such as how to run a jar test on raw water to determine the appropriate coagulant dose under variable water quality conditions. Such training opportunities mean that water system operators learn about new techniques to keep their systems in good working order as well as learn how to meet new regulatory requirements. Training also helps operators maintain the necessary certifications to properly operate a variety of treatment technologies, noting that advanced water treatment technologies take a highly trained operator with a suite of math, water chemistry, mechanical, and computer skills to operator correctly. Through these state programs, operators may also improve their range of knowledge and take on greater responsibilities.

Set-asides also give states the flexibility to find ways to work with individual water systems to enhance and maintain their technical, managerial, and financial capabilities. This capacity development program allows states to provide support for struggling systems as a means to attain and maintain compliance without resorting to financial penalties or other enforcement mechanisms. While we tend to group systems into broad challenge categories –disinterested owners, no certified operator, inadequate rates – the reality is that each system presents a unique set of circumstances. Although technical proficiency is often the most obvious key to compliance, efficient management and effective financial strategies are equally critical components of a well-run water system. Capacity development, through the set-asides, allows the state drinking water program to respond to those unique challenges and fashion an effective and achievable solution for that system.

Source water and wellhead protection programs also make use of set-asides. As a first step, statewide source water assessments were developed to provide the framework for local utility-based source water protection programs. These programs can help systems avoid additional costs for increasing treatment capacity and avert the need to install advanced treatment technologies. Other uses include support for public water systems to update their assessments and develop and implement protection plans and voluntary, incentive-based actions such as agricultural resource and livestock management, land acquisition, and conservation easements that help protect source water quality and reduce nonpoint source pollution. As well, the set-asides are also used to implement stormwater management projects, abandoned well programs, and efforts to address malfunctioning septic systems to reduce the infiltration of contaminants into underground sources of drinking water.

While the above outlines many of the principal uses for set-asides, there are other ways that set-asides offer new opportunities for water systems to improve their capabilities. Set-asides also provide training on new science-based regulations, allow technical assistance to help systems understand and meet new rule requirements, and provide training for communications protocols such as Consumer Confidence Reports and enhanced Public Notification. States also have been enthusiastic partners with the EPA in bringing the concept of asset management to smaller water systems through training and technical assistance. Set-asides also support efforts such as developing performance-based training strategies and sharing area wide optimization protocols.

More recently, state drinking water programs, in concert with EPA, have been looking at new ways to encourage water systems to consider a range of low or no cost options to enhance their capabilities. They include a range of tools and resources such as shared purchasing, shared spare parts and back-up equipment, broader use of contract operations, contracted services (meter reading, payroll, billing, etc.), water line extensions (where feasible), system consolidation, formation of a regional water system, or purchase of an adjacent system. Loosely termed "partnerships," these options can be as simple as sharing a backhoe or as complex as merging with a neighboring system. States are encouraging systems to evaluate which, if any, of the range of options may improve their operations and better position them to protect public health. States are prepared to work with these systems should they decide to modify their management or operations.

As an example of how states use these set-asides in a proactive manner, I'd like to tell you about the Stockton Water System in my own state of Pennsylvania. Stockton is a very small (43 homes) community. Although its water system may have been in operation for several years, the original homeowners/managers/operators had moved away or passed away and Stockton was operating as an untreated, unfiltered, and unpermitted surface water system. The state "discovered" the system in 2014 because of customer complaints and found several pathogens in the water – *E. coli, Giardia*, and Salmonella just to name a few. While the violations continued to mount, traditional strategies or enforcement actions would have been of little practical use to the residents of Stockton. They needed a different kind of help to regain confidence in their drinking water and know that their water was safe to drink.

Pennsylvania has developed two capacity development-based initiatives that came into play for Stockton. The first, the Capability Enhancement Facilitator Program, provided an initial assessment of TMF capability and also provided onsite technical assistance and education to help Stockton and local and state representatives understand the challenges. The second, Pennsylvania's Professional Engineering Services Program, helped with feasibility studies and design work to find the best approach to return Stockton to compliance. These two initiatives came together with the state's DWSRF program – PENNVEST. PENNVEST is the states SRF funding agency.

The three state-based programs worked collaboratively to identify what the state calls a "White Knight," a willing, well run partner, in this case the nearby Hazelton City Authority. Hazelton agreed to work with Stockton, make the DWSRF application, extend water service, and replace the existing distribution system in Stockton and keep water rates at an affordable $35.50 per month. Total project cost? $2.21 million underwritten by PENNVEST. Stockton now has a safe, reliable supply of drinking water and Hazelton now has 43 new customers.

Solutions such as this one would not be possible without the availability of set-asides to support water system challenges; providing an achievable path to a DWSRF loan; and restoring public health protection. In short, drinking water systems, and ultimately the public through increased public health protection, are the direct beneficiaries of the work accomplished through the DWSRF set-asides.

THE PWSS PROGRAM

The Public Water System Supervision or PWSS program forms the critical core of all of our public health protection efforts. This program provides the means for state drinking water programs to ensure that all public water systems – large and small communities, schools, child care facilities, restaurants, places of business, highway rest stops, campgrounds – provide a reliable and safe supply of water that is available for all thirsty Americans. For more than 40 years, states have willingly accepted these responsibilities. In recent years, state drinking water programs have accepted additional responsibilities in water system security and resiliency that include working with all public water systems to ensure that critical drinking water infrastructure is protected, including cyber security; that plans are in place to respond to both natural and manmade disasters; and that communities are better positioned to support both physical and economic resilience in times of crisis.

More Work is Needed

From a public health perspective, 93% compliance with health-based standards is not optimal. State drinking water programs continue to strive toward a higher national goal for public health protection. The ability to support our water systems is essential to success for our communities. However, state drinking water programs are extremely hard pressed financially and the funding gap continues to grow. States must accomplish all the above-described activities – and take on new responsibilities – in the context of a challenging economic climate. State-provided funding has historically compensated for inadequate Federal funding, but state budgets have been less able to bridge this funding gap in recent years. State drinking water programs have often been expected to do more with less and states have always responded with commitment and integrity, but they are currently stretched to the breaking point. Insufficient funding support for these critical programs increases the likelihood of contamination events that put the public's health at risk. We do not want to see another Charleston, WV or Toledo, OH, or Flint, MI.

State drinking water programs want to broaden their support for water systems. They want to find the best solutions for the greatest number of system needs. They want to be able to help water systems sustain their abilities to provide reliable, safe water supplies at a reasonable cost to their customers. This is the public health protection goal for the drinking water community.

To achieve this goal, however, states need Congressional support. For the past four years, state PWSS programs have been flat funded at $101.9 million. The DWSRF has seen decreased funding. In FYs 14 and 15, $906.8 million was awarded for the infrastructure loan program but FYs 16 and 17 saw the award decrease to $863.2 million. These essential public health programs with well-documented needs and successes must be fully supported, even in these economically challenged times.

ASDWA recommends that the PWSS program be funded at $200 million to allow states to continue their core programmatic work with water systems to ensure safe drinking water. Additionally, ASDWA recommends that the DWSRF be funded at $1 billion. This funding level will provide needed funding to make additional awards for infrastructure improvements, work with distressed communities in need, support small systems that need assistance to sustain their capabilities, and continue to provide training and technical assistance on the wide array of rules and regulations designed to protect public health. ASDWA also recommends that funding continue for WIFIA, as this new funding program shows promise for increasing the level of infrastructure investments.

SUMMARY

In summary, the 1996 SDWA Amendments offered the water community a promise of enhanced public health protection through a framework of traditional and proactive collaborations between state drinking water programs and the water systems they oversee. Much progress has been made in these efforts and more work is needed to protect public health and maintain the economic health of our communities. State drinking water programs work in partnership at the Federal level with EPA and other Federal health and environmental programs.

Equally important are our local partners – the water utilities themselves. While most Americans receive, by volume, their drinking water from medium and large water utilities, the vast number of the 152,000 public water systems in the United States are small (more than 90% of all community water systems serve less than 10,000 people). Innovative tools and resources are needed to increase compliance for this large cohort of small systems. The level of effort required to support and sustain each of these systems requires trust and collaboration and a willingness to partner. State drinking water programs are committed to fulfilling the promise of the 1996 SDWA Amendments.

Mr. Shimkus. I thank the lady.

The chair now recognizes Mr. Kurt Vause, who gets the longest traveling award for getting here, Special Projects Director at Anchorage Water and Wastewater Utility, on behalf of the American Water Works Association.

You are recognized for 5 minutes. Welcome.

STATEMENT OF KURT VAUSE

Mr. VAUSE. Good morning, Chairman Shimkus, and members of the subcommittee. My name is Kurt Vause. I am the Special Projects Director for the Anchorage Water and Wastewater Utility from Anchorage, Alaska. I also serve as the Chair of the Water Utility Council and Acting Chair of the Asset Management Committee of the American Water Works Association.

We deeply appreciate this opportunity to offer the viewpoints and experiences of drinking water providers to the important deliberations and decisions of this committee.

The discussion draft of drinking water legislation this subcommittee is considering is a good step towards addressing the Nation's needs, to reinvest in its water infrastructure, and towards addressing other needs as well. I would like to briefly address three topics.

First, providing safe drinking water to communities requires a complex mix of engineering, capital investment, management, science, community engagement, and regulatory resources. This complexity makes it particularly difficult for many small systems to remain in compliance in regulation and maintain their infrastructure.

Options to help address these challenges include partnerships or regionalization to share resources among these systems, many who serve small systems and communities. Regionalization or partnerships encompass anything from physical connections to shared management, engineering, operations, and purchasing resources.

When a compliant utility absorbs or merges a noncompliant utility, that newly formed utility faces a regulatory compliance challenge. The SDWA ought to provide a finite grace period for the newly merged system to come into compliance with regulation. Whether a utility has explored consolidation should become one of the factors weighted in ranking SRF loans or in evaluating compliance options.

Second, all utilities manage their assets, but the practice we now formally call asset management is more scientific and focused. The goal of infrastructure asset management is to meet a required level of service in the most cost-effective manner at an acceptable level of risk through the management of assets for present and future customers.

We do not believe a specific level of asset management practice should be mandated, because that would put Congress or a regulatory agency in the business of defining asset management objectives. Utilities vary too greatly in strategic objectives, size, type of assets, geography, climate, source waters, type of treatment and distribution for a Federal definition to be practical.

Professional organizations such as AWWA are making education and asset management practice an ongoing part of our educational efforts for members. For example, AWWA's upcoming annual conference. Our Asset Management Committee has developed a track of sessions on project infrastructure and asset management with five individual sessions containing 27 separate presentations.

We also believe there is a role States can play in similar efforts through the maintenance of the PWSS supervision grants. We urge Congress to maintain PWSS funding for fiscal year 2018 at no less the current authorization levels.

Third, as we have said before to Congress, local rates and charges have been and will likely always be the backbone of local water system finance. However, when major infrastructure projects require either to comply with regulations or replace aging infrastructure, there is a need for a quicker, larger infusion of cash than those rates and charges typically provide.

This is where the toolbox of utility finance comes into play. This spring, AWWA cosigned a two-page summary of how the Federal Government can assist water utilities in financing these challenges. The highlights of that were: Number one, preserve the tax-exempt status of municipal bonds; two, provide fully authorized funding for the Water Infrastructure Finance and Innovation Act, known as WIFIA, at $45 million for fiscal year 2018; three, double appropriations for the drinking water and wastewater SRF programs; and four, remove the annual volume caps on private activity bonds for water infrastructure projects.

We realize appropriations come from the Appropriations Committees, but we seek your support in funding with these panels.

This concludes my remarks to the subcommittee. We also look forward to continuing dialogue with this panel after this hearing.

[The prepared statement of Mr. Vause follows:]

Funding, Management, and Compliance Assistance under the Safe Drinking Water Act
Presented by Kurt Vause
Special Projects Director
Anchorage Water & Wastewater Utility & Chair, Water Utility Council
American Water Works Association
Before the House Subcommittee on the Environment
May 19, 2017

Good morning, Chairman Shimkus and members of the subcommittee. My name is Kurt Vause and I am Special Projects Director for the Anchorage Water and Wastewater Utility in Anchorage, Alaska. I also serve as chair of the Water Utility Council and acting chair of the Asset Management Committee of the American Water Works Association, on whose behalf I am speaking today. We deeply appreciate this opportunity to offer the viewpoints and experiences of drinking water providers to the important deliberations and decisions of this committee.

The Safe Drinking Water Act last saw significant amendment in 1996. That bill was an important improvement over the previous act as it created a very useful finance tool, the state revolving loan program, and set down a data-informed, methodical process for setting new regulations and revising existing ones. The last point is very significant. The Stage 2 Disinfection By-Products Rule and Enhanced Surface Water Treatment Rule were major rulemaking efforts and have resulted in significant investments to upgrade treatment plants across the country. In the coming months, we expect to see a revised Lead and Copper Rule that will trigger important changes in the way communities address lead exposure. However, an updating of the 1996 Amendments to the SDWA is overdue. Our 2012 report, “Buried No Longer: Confronting America’s Water Infrastructure Challenge” pointed out that this nation must spend $1 trillion on drinking water infrastructure in the next 25 years to maintain our current levels of service. Based on our past observations, the cost of maintaining wastewater infrastructure are about equal. The discussion draft of drinking water legislation the subcommittee is considering takes a good first step in that direction. I will address certain features in this early draft and some additional issues.

CONSOLIDATION, PARTNERSHIPS OR REGIONALIZATION

Providing safe drinking water to communities requires a complex mix of engineering, capital investment, management, science, community engagement, and regulatory resources. This complexity makes it particularly difficult for many small systems to remain in compliance wiith regulation and maintain their infrastructure. Some small systems are finding it increasingly difficult to remain in compliance with regulations and remain fiscally sustainable. One option to help address the sustainability challenge leverages consolidation or regionalization to share resources among these systems, many whom serve small communities. Regionalization for water utilities encompasses anything from physical connection to shared management, engineering, operations or purchasing resources.

States at times encourage consolidation of these smaller, struggling systems into neighboring larger, stronger water utilities. The larger utility faces a regulatory compliance burden in these merger or acquisition situations when the challenged system is not in compliance with regulations. By merging, the larger utility inherits the compliance challenge, and status, of the utility it means to serve. The SDWA ought to provide a grace period for the newly merged system to come into compliance with regulation. We understand this would have to be a finite period of time and would be happy to sit down with the committee and work out more details.

Some form of consolidation or regionalization should be one of the options a water system explores when it faces regulatory compliance or financial challenges. This exploration could become one of the factors weighed in ranking SRF loans or in bringing a system into compliance. The authorizing language for the state revolving loan fund prohibits the use of an SRF loan to "finance the expansion of any public water system in anticipation of future population growth." This effectively prohibits accessing an SRF loan until after a community has already grown. The rapid growth of communities in suburbs, the Sunbelt, the West, and even some city centers, already makes keeping up with infrastructure needs a challenge without the expansion prohibition of SRF. Drinking water and wastewater pipes, as well as roadways and sidewalks, must be built to meet the growing needs of a community, or in lockstep with rehabilitation efforts. We understand that the original intent of the language of the SRF was to prohibit use of this funding to support reckless sprawl. However, population trends, including infill, brownfield reclamation and urbanization make this provision obsolete for many communities.

The law could be improved by making it clear that using the SRF to help finance projects of consolidation for efficiency of operations and regulatory compliance does not violate the anti-sprawl provision. It should also give more leeway to utilities that clearly see future growth in certain areas near their current service areas.

ASSET MANAGEMENT

All utilities manage their assets, but the practice we now formally call asset management is more scientific and focused. The goal of infrastructure asset management is to meet a required level of service, in the most cost effective manner, at an acceptable level of risk, through the management of assets for present and future customers. (AWWA, 2015). Advanced asset management practice helps a utility understand the state of what assets it has, the required service levels assets are to provide, the risk of asset failure to achieving utility objectives, and what operations and maintenance strategies are best to use. It is matched with the development of a long-range financial plan to finance and fund utility operations so together, the right assets are available at the right time for the right price. This knowledge helps utilities get the most out of the dollars invested and meet required service standards.

We do not believe a specific level of asset management practice should be mandated because that would put Congress or a regulatory agency in the business of defining asset management practices. Utilities vary too greatly in strategic objectives, size, types of assets, geography, climate, source waters, types of water treatment and distribution, etc., for a federal definition to be practical. Professional organizations such as AWWA are making education in asset management practice an ongoing part of our educational efforts for members. For example, for AWWA's annual conference to be held this coming June, I helped develop a track of sessions on project infrastructure and asset management with five individual sessions containing 27 separate presentations. For our Water Infrastructure Conference in Houston in the fall, AWWA's Asset Management committee was asked to assemble a hands-on session for developing asset management plans.We have a web page dedicated to asset management that provides access to publications, journal articles, and similar resources. We believe in educating water providers and related professionals about leading asset management practices and will continue our outreach efforts in this field.

PUBLIC WATER SYSTEM SUPERVISION (PWSS) GRANTS

Last month, AWWA cosigned a letter to congressional appropriators urging that PWSS grants not be cut in the fiscal year 2018 budget, as was proposed in the president's budget. We explained, "State drinking water programs use PWSS funds to ensure that water utilities have the information, technology, and capabilities to meet their mandated regulatory responsibilities – an essential component of public health protection.

"Utilizing the PWSS grants, these state programs provide educational programs, training and technical assistance where needed. In other words, the PWSS grant program provides the means for states to work with drinking water utilities to ensure that American citizens can turn on their taps with confidence that the water is both safe to drink and available in adequate quantities.

"PWSS funds are distributed to the states, five territories, and the Navajo Nation to provide oversight of approximately 151,000 public water systems; assist in their understanding of their regulatory responsibilities; and assist in consistent compliance and enforcement of drinking water regulations, particularly where public health may be threatened."

Cosigning the letter were the Association of Metropolitan Water Agencies, Association of State Drinking Water Administrators and the National Association of Water Companies. It is Attachment A to this testimony.

SRF Enhancement

We addressed the subcommittee earlier in the year about areas for exploring improvement in the state revolving loan fund program. We will reiterate that the application process seems to widely vary from state to state. We encourage the U.S. Environmental Protection Agency to convene SRF stakeholders to develop educational materials to help guide states streamline and normalize the application and loan capitalization process. Right now, we see a once-a-year snapshot of undisbursed SRF balances in each state. The report we saw from June 2016 showed states with everywhere from 2 to 38 percent of their SRF capitalization funds undispersed at the end of the states' fiscal years. That annual snapshot may not be be fairly portraying how efficiently states are moving money or it may be showing where help is needed to get SRF loans out the door. We just don't have the data to know either way. We urge that states be required to provide quarterly snapshots of undisbursed balances to we know where help is needed.

SRF loans require recipients to track compliance with state and federal goals for minority, women and/or disadvanted business enterprise participation. To comply with these goals, different programs of various primacy agencies can stipulate different methods of tracking. This is another example of where we encourage the U.S. Environmental Protection Agency to convene SRF stakeholders to develop educational materials to help guide states in order to streamline recipients administration of loans.

Another enhancement is added flexibility in repayment terms of SRF loans. To some communities, the terms of repayment will necessarily lead to a limited use of SRF financing of critical infrastructure needs. Adding more flexibility in repayments, such as longer periods for repayment of principal and interest on loans, not to exceed the useful lives of assets acquired, offers states another way to enhance affordability.

The SRF currently requires compliance with Davis-Bacon and Buy America laws and proof of cross-cutting compliance with other environmental laws. Altogether, this not only raises the burden of application for an SRF loan – particularly for smaller systems – but exposes the utility receiving the loan to additional legal hazards. A number of states and municipalities have their own Davis-Bacon-like or Buy America or environmental cross-cutter laws. In such states, the federal requirements are a redundancy and still require their own documentation. We applaud efforts to try and streamline the cross-cutter requirements in the discussion draft. We encourage the committee to look at other opportunities to streamline similar requirements. For example, there is a waiver available from Buy America requirements if the cost of domestic materials causes the cost of the entire project to increase by 25 percent. This is an unrealistic requirement as materials alone often are less than 25 of a total project's costs. We urge that this be changed to make the waiver available if the domestically produced material costs in question themselves are more than 25 percent greater than materials meeting the same quality and performance requirements.

SOURCE WATER PROTECTION

The necessity of protecting our source waters was dramatically illustrated in August 2014 when the Toledo, Ohio water system had to shut down because of harmful algal blooms in Lake Erie.

AWWA and other water associations believe strongly that it is better to prevent contaminants from entering a watershed than to treat them after they have entered water supplies. That is why we have, for example, ramped up efforts to educate water utilities about partnership programs at the U.S. Department of Agriculture in which utilities work out source water solutions in a cooperative manner with upstream farmers and ranchers.

We note that the discussion draft would allow up to 10 percent of the annual SRF capitalization grants to a state to be used for source water delineations, assessments or updates. We note that already, states are allowed to use up to 4 percent of the capitalization grants to administer the SRF and another 27 percent for other purposes. In this era when we are trying hard to reinvest in our nation's water infrastructure, we question this diversion. We would definitely want to see such diversions capped to a finite number of years, such as fours years, as it was in the 1996 Amendments.

WATER INFRASTRUCTURE FINANCE

As we have said before to Congress, local rates and charges have been, and will likely always be the backbone of local water system finance. However, when major infrastructure projects are required, either to comply with regulations or to replace aging infrastructure, there is a need for a quicker, larger infusion of cash than those rates and charges can provide. That is where the toolbox of utility finance comes into play. This spring AWWA cosigned a two-page summary of how the federal government can assist water utilities in finance challenges. The highlights are as follows:

1. Preserve the tax-exempt status of municipal bonds.
2. Provide fully authorized funding for the Water Infrastructure Finance and Innovation Act (WIFIA).
3. Double appropriations for the drinking water and wastewater SRF programs.
4. Remove the annual volume caps on private activity bonds for water infrastructure projects.

Note that earlier in this testimony, we recommended improved tracking of SRF capitalization grants. We urge that Congress and EPA implement measures such as quarterly reporting of undisbursed SRF funds before providing additional SRF funds. The committee is already familiar with the value of the SRF program, particularly for water systems with the greatest compliances challenges. WIFIA is a relatively new program, but its potential value for rehabilitating the nation's water infrastructure was illustrated dramatically this spring. Congress appropriated funds for WIFIA to begin making loans last December. Applying for a WIFIA loan is a two-step process. First a utility or community sends a letter of interest to EPA. That triggers a dialogue between the agency and the utility or community. Then if the agency sees that the utility or community is likely to qualify for a WIFIA loan, it encourages the utility or community to file a formal application.

EPA accepted the first round of letters of interest until midnight April 10. It received 43 letters of interest for drinking water, wastewater and stormwater projects. Congress appropriated enough money for WIFIA to award about $1 billion in loans. The letters of interested received in April sought about $6 billion in WIFIA loans, and because WIFIA only funds 49 percent of a project's costs, that means those letters of interest were for about $12 billion in water infrastructure work. We are grateful for the funds Congress appropriated in December and for the additional $10 million appropriated in the recent continuing resolution, and we urge Congress to appropriate the fully authorized $45 million for FY2018. WIFIA represents a great investment for the federal government since it is strictly a loan program with no grants. Funds supporting infrastructure projects come back to the Treasury. Modeled after the successful transportation program, TIFIA, WIFIA leverages appropriations to maximize investment. The credit history of water utilities supports WIFIA's ability to provide a leverage ratio of up to 1:65 according to congressional estimates. A fully authorized FY2018 WIFIA would support nearly $3 billion in needed infrastructure investment.

Cosigning the two-pager on finance with AWWA were the Association of Metropolitan Water Agencies, the National Association of Clean Water Agencies, the Water Environment Federation, the U.S. Water Alliance, the Water Environment Research

Foundation, the Water Research Foundation, WateReuse and the Water and Wastewater Equipment Manufacturers Association. It is Attachment B. We realize that the tax code and actual appropriations are outside the jurisdiction of this committee, but we do urge you to contact your collcagues on the relevant committees in support of these policies and funding.

INTEGRATED PLANNING

AWWA has taken notice of work by various members of Congress to help provide states and municipalities with greater flexibility to prioritize and more effectively manage obligations under the Clean Water Act (CWA). In fact, just yesterday, the House Transportation and Infrastructure Subcommittee on Water Resources and Environment held a hearing on "Improving Water Quality through Integrated Planning." This hearing examined the difficulties that communities face in meeting the regulatory requirements of the CWA given dwindling resources, as well as codifying the 2012 Integrated Planning Framework developed by EPA in order to help communities meet their regulatory obligations. AWWA is pleased with this development, and would like to urge this subcommittee to expand that work and bring the drinking water sector into the integrated planning process. Communities and municipalities don't look at their regulatory obligations in a vacuum, and must view water holistically. AWWA recommends Congress include drinking water requirements containined with the 1996 amendments of SDWA in any integrated planning framework to give communities across the country the flexibility to more effectively meet their regulatory obligations, while also better protecting public health.

This concludes my remarks, and I will be happy to take questions from the subcommittee. We also look forward to continued dialogue with this panel after this hearing.

WHAT IS THE AMERICAN WATER WORKS ASSOCIATION?

The American Water Works Association (AWWA) is an international, nonprofit, scientific and educational society dedicated to providing total water solutions and assuring the effective management of water. Founded in 1881, the association is the largest organization of water professionals in the world.

Our membership includes more than 4,000 utilities that supply roughly 80 percent of the nation's drinking water and treat almost half of the nation's wastewater. Our 50,000 members represent the full spectrum of the water community: public water and wastewater systems, environmental advocates, scientists, academicians, and others who hold a genuine interest in water, our most important resource. AWWA unites the diverse water community to advance public health, safety, the economy, and the environment.

April 25, 2017

The Honorable Ken Calvert
Chair

The Honorable Betty McCollum,
Ranking Member
House Appropriations Subcommittee on Interior, Environment and Related Agencies

The Honorable Lisa Murkowski
Chair

The Honorable Tom Udall
Ranking Member
Senate Appropriations Subcommittee on Interior, Environment and Related Agencies
United States Congress
Washington, DC

Dear Chairman Calvert, Ranking Member McCollum, Chairwoman Murkowski, Ranking Member Udall,

We, the undersigned organizations, are all dedicated to protecting the public health and economic health of communities across the nation through the provision of safe drinking water. We urge Congress to help in this by sustaining funding that allows state drinking water programs to achieve these goals through the Public Water System Supervision (PWSS) grant program, as codified in the Safe Drinking Water Act.

We all understand the escalating need to reinvest in the country's water infrastructure. We endorse actions such as increasing funding for the drinking water state revolving loan fund program (DWSRF) and the Water Infrastructure Finance and Innovation Act (WIFIA) program. However, building or renewing infrastructure alone is not the solution for protecting the health of our citizens and the economic health of our communities.

State drinking water programs use PWSS funds to ensure that water utilities have the information, technology, and capabilities to meet their mandated regulatory responsibilities – an essential component of public health protection. Utilizing the PWSS grants, these state programs provide educational programs, training and technical assistance where needed. In other words, the PWSS grant program provides the means for states to work with drinking water utilities to ensure that American citizens can turn on their taps with confidence that the water is both safe to drink and available in adequate quantities.

According to a 2013 resource-needs analysis of state drinking water programs, "...even as resource needs are increasing, the funding availability to support the state drinking water programs in their mission has stagnated...if funding continues at current levels, states will not have adequate funding to support their minimum base programs over the next ten years." (Insufficient Resources for State Drinking Water Programs Threaten Public Health: An Analysis of State Drinking Water Programs' Resources and Needs, December 2013)

Federal funding for this vital program has essentially remained flat for the past several years at $101.9 million. However, the need for state drinking water programs to perform mission-critical functions has never been more important. Funding for this program must not be decreased, particularly given the ever-increasing responsibilities states are taking on in public health protection. Without robust funding, we are shortchanging ourselves.

PWSS funds are distributed to the states, five territories, and the Navajo Nation to provide oversight of approximately 151,000 public water systems; assist in their understanding of their regulatory responsibilities; and assist in consistent compliance and enforcement of drinking water regulations, particularly where public health may be threatened. More than 90% of the U.S. population receives water for bathing, cooking, and drinking from a public water system – overseen by state drinking water program personnel.

Having economically vibrant communities, healthy citizens, a productive workforce, and sound businesses depends on a safe and reliable supply of drinking water. Through the PWSS program, state drinking water programs support the water utilities that are essential to these goals.

Please maintain funding for the PWSS program at least at current levels.

Thank you,
G. Tracy Mehan, III
Executive Director Government Affairs
American Water Works Association

Diane VanDe Hei
Chief Executive Officer
Association of Metropolitan Water Agencies

Michael Deane
Executive Director
National Association of Water Companies

Alan Roberson
Executive Director
Association of State Drinking Water Administrators

ELEVATE WATER AS A NATIONAL PRIORIT

America's Economic Future Depends on Safe and Clean Water

America's future economic strength depends on investments made today in water infrastructure. These investments create jobs and support the economy. Consider these facts: Every $1 invested in water and wastewater infrastructure increases long-term GDP by $6.35; each job created in water and wastewater leads to 3.68 jobs in the national economy; over $86 billion annually is spent on water-related sports activities. Studies also show that the US economy would stand to gain over $200 billion in annual economic activity and 1.3 million jobs over a 10-year period by meeting its water infrastructure needs. But, without this investment, breakdowns in water supply, treatment and wastewater capacity are projected to cost manufacturers and other businesses over $7.5 trillion in lost sales and $4.1 trillion in lost GDP from 2011 to 2040.

America's Quality of Life Depends on Safe and Clean Water

Well-functioning water and wastewater systems, and the research efforts to support them, are critically important to America's quality of life. Past investments in drinking water, wastewater and stormwater infrastructure have left America with some of the best drinking water in the world, while providing our children with safe water for swimming and bathing, and our cities and towns with opportunities to revitalize waterfronts to support new businesses, residences, and recreational activities. However, investment in water, wastewater and stormwater infrastructure and research has failed to keep pace with maintenance demands and emerging hydrological threats, putting our quality of life gains at risk.

Federal Investment Ensures Safe and Clean Water

Since enactment of the Clean Water Act in 1972 and the Safe Drinking Water Act in 1974, Congress has supported a strong federal funding partnership with States and local ratepayers to pay for this critical infrastructure through:

- Investments in the Drinking and Clean Water State Revolving Funds, which return over $.93 to the Federal Treasury for every $1 invested;
- Tax-exempt municipal bonds, which financed nearly $38 billion in water and wastewater infrastructure in 2016; and,

- WIFIA, the Water Infrastructure Finance and Innovation Act, which has the potential to leverage over $60 for every $1 invested in major water and wastewater projects.

Yet EPA estimates that America's water and wastewater infrastructure requires more than $650 billion worth of investment over the next 20 years just to maintain cur- rent levels of service, and independent estimates place this figure over $1 trillion. While local ratepayers will shoulder much of this burden, all levels of government must be part of the solution.

Mr. Shimkus. The chair thanks the gentleman.

The chair now recognizes Ms. Lynn Thorp, National Campaigns Director at Clean Water Action.

You are recognized for 5 minutes.

STATEMENT OF LYNN THORP

Ms. THORP. Thank you.

Good morning, Chairman Shimkus, Ranking Member Tonko, and members of the subcommittee. My name is Lynn Thorp. I am National Campaigns Director at Clean Water Action. We are a national organization with 1 million members working in 15 states on health and environmental projects with an emphasis on drinking water issues.

Thank you for the opportunity to provide comments on the Drinking Water System Improvement Act. Recent high profile events have highlighted the importance of infrastructure investment, effective system operation, and source water protection. From the drinking water crisis in Flint, Michigan, to the leaking chemical storage tank that contaminated the Elk River in West Virginia, we have seen how taking drinking water for granted can lead to public health risks and economic disruption of entire communities.

Our approach to meeting 21st century drinking water challenges needs to be a holistic one. It should include not only increased investment in infrastructure, but also sufficient resources for effective oversight of Safe Drinking Water Act compliance by Federal and state primacy partners, more funding for research and innovation, more attention to keeping drinking water sources clean, and a vision for how we want our drinking water systems to look in the second half of the 21st century.

You can see some ideas about that in the testimony from the witnesses we have heard from already this morning and in the 2016 U.S. Environmental Protection Agency Drinking Water Action Plan.

We do hope this subcommittee will consider provisions in the Safe Drinking Water Act Amendments of 2017, H.R. 1068, introduced by Representatives Tonko and Pallone earlier this year. Transparency, how we determine which contaminates to regulate, climate resiliency and drought, threats to drinking water from oil and gas and other activities, water efficiency, and technology innovation are all important if we are to maintain a high quality of drinking water and healthy water systems.

We support Drinking Water State Revolving Fund authorizations commensurate with those proposed in the AQUA Act mentioned earlier today, which would authorize over $3 billion in fiscal year 2018, and increase thereafter reaching $5.5 billion on fiscal year 2022.

AWWA, the American Society of Civil Engineers, and EPA have repeatedly found investment needs orders of magnitude greater than those authorizations I have mentioned. Ambitious authorizations signal a commitment to clean drinking water and are a reasonable contribution to the mix of funding sources available to drinking water systems.

We also support increased authorizations for Public Water Systems Supervision grants. The Association of State Drinking Water Administrators has estimated the gap in needs between current funding and comprehensive State programs to be $300 million or more annually. As noted earlier, bridging this gap will increase public health protection and support sustainable drinking water systems.

Drinking Water State Revolving Fund dollars can be spent on numerous activities that support those goals: pipe replacement, treatment upgrades, source water protection, improvements for storage, and system restructuring and consolidation. We want to highlight just two of those here as examples: pipe repair and replacement and source water protection.

As you know, EPA estimates we may have between some 6 1A½ or even more than 10 million lead service lines or partial lead service lines in the United States. Lead is a highly poisonous metal, and children under 6 are most at risk. Increased investment can help more communities move sooner to full lead service line replacement.

American Society of Civil Engineers also estimates there are over 240,000 water main breaks each year due to deteriorating and poorly maintained pipes. As you probably know, just this week, a pipe from 1860, a water main broke right in Northwest, D.C. We lose water through leaks in mains and service lines as well, and these disruptions threaten public health, allowing pathogens to get into the pipes and, of course, lead to loss of treated water. Some estimates say up to 18 percent of treated water, which is a valuable commodity, if you will.

So shoring up our underground drinking water infrastructure not only protects public health, reduces lost revenue for drinking water systems, but also leads to less disruption, like we saw in parts of D.C. just this week.

We can also use Drinking Water State Revolving Funds for source water protection, and many communities are using innovative strategies in this area. The return on investment there is clear in terms of public health protection. And the EPA estimates that every dollar spent on protecting a drinking water source saves $27 in drinking water treatment.

I just want to close by noting that EPA programs are fundamental to the success of state programs and water systems. So increased state revolving fund investment won't be as effective if at the same time EPA lacks staffing and funding for oversight, enforcement, research, development of contaminant standards, support for small systems, and other critical activities.

We urge subcommittee members to oppose cuts in EPA funding as well as rollbacks of health and environmental protections that would put our Nation's drinking water sources at risk of contamination.

Thank you for the opportunity to provide these comments.

[The prepared statement of Ms. Thorp follows:]

Testimony of Lynn W. Thorp
National Campaigns Director
Clean Water Action
Before the U.S. House of Representatives Energy and Commerce Committee
Subcommittee on the Environment

Good morning. I am Lynn Thorp, National Campaigns Director at Clean Water Action. We appreciate the opportunity to provide testimony at today's hearing. Clean Water Action is a national organization working in 15 states on a wide range of environmental and health issues. Our work includes a focus on Safe Drinking Water Act implementation and on protecting drinking water sources through upstream pollution prevention programs. I have worked at Clean Water Action for 18 years. I have served two terms on the National Drinking Water Advisory Council (NDWAC), which advises the U.S. Environmental Protection Agency (EPA) on drinking water issues. I have also served on Federal Advisory Committees and NDWAC Work Groups to consider major SDWA implementation activities including revisions to the Lead and Copper Rule and the Total Coliform Rule and development of the Contaminant Candidate List process.

Thank you for the opportunity to provide comments on the *Drinking Water System Improvement Act* and on the critical issues involved in Safe Drinking Water Act compliance and on how federal requirements and support can help ensure that our nation's drinking water systems are doing the best possible job of protecting, treating and distributing drinking water.

Over the last several years, high profile drinking water contamination events focused renewed attention on drinking water and highlighted the importance of infrastructure investment and source water protection. From the drinking water crisis in Flint, Michigan to the leaking tank that contaminated the Elk River in West Virginia, we have seen how taking drinking water for granted can lead to public health risk and economic disruption of entire communities. There is a critical need to invest in our nation's drinking water infrastructure and in other activities that support ensuring Safe Drinking Water Act compliance and sustainable management of drinking water systems. Our approach to meeting 21st century drinking water challenges needs to be a holistic one. It needs to include not only increased investment in infrastructure through programs including the Drinking Water State Revolving Fund, but more effective oversight of Safe Drinking Water Act compliance by federal and state primacy partners, more funding for research and dedication to innovation, more attention to keeping contamination out of our nation's drinking water sources, and a vision for how we want our drinking water systems to look in the second half of the 21st century. Some valuable ideas for this holistic approach are reflected in the U.S. Environmental Protection Agency's (EPA) 2016 *Drinking Water Action Plan.*

THE FUTURE OF THE SAFE DRINKING WATER ACT

The Safe Drinking Water Act Amendments of 2017, H.R. 1068, introduced by Representatives Tonko and Pallone earlier this year, includes provisions around critical issues for the future of clean drinking water in the United States. We hope that as the Subcommittee considers *The Drinking Water Improvement Act* that it will consider these provisions, including those around lead, drinking water in schools, climate resiliency and drought, oil and gas threats to drinking water, monitoring technology research and system restructuring. Our comments today are focused primarily on increased investment through the Drinking Water State Revolving Fund program.

INCREASE AUTHORIZATIONS FOR THE DRINKING WATER STATE REVOLVING FUND PROGRAM

We need to substantially increase investment in drinking water through the Drinking Water State Revolving Fund. The discussion draft provided for today's hearing did not include details on proposed authorizations. We support funding levels commensurate with those proposed in the *"AQUA Act," Title IV of H.R. 1068*, which would authorize over $3 billion in fiscal year 2018 and increase thereafter reaching $5.5 billion in fiscal year 2022. The American Water Works Association, the American Society of Civil Engineers, and EPA have repeatedly determined drinking water investment needs to be orders of magnitude larger than these proposed levels. These levels of authorization for the Drinking Water State Revolving Fund represent a quite reasonable contribution from one source of federal investment in clean drinking water, a goal resoundingly supported by the public.

The return on investment for increased Drinking Water State Revolving Fund authorizations will be experienced in more public health protection, stronger local economies, and more sustainable drinking water systems. Drinking Water State Revolving Fund dollars can be spent on numerous activities that support these goals, including pipe replacement, treatment upgrades, source water protection, storage improvements and system restructuring. We note two examples – pipe repair and replacement and source water protection.

DRINKING WATER SRFS - PIPE REPLACEMENT – LEAD SERVICE LINES AND LEAKS

Drinking Water State Revolving Funds can be used for water main and service line replacement. EPA estimates that there are 6.5 million to perhaps over 10 million service lines made at least partially of lead in the United States. Many communities are interested in getting these service lines out of the ground to eliminate the largest source of lead in contact with water. Lead is a highly poisonous metal and can affect almost every organ in the body and the nervous system. Children under six are most at risk from lead poisoning. Even low levels of lead exposure have been found to permanently reduce cognitive ability and cause hyperactivity in children. Full lead service line replacement projects can not and will not be financed solely by Drinking Water State Revolving Fund investment, but increased authorizations can help more communities move sooner to full lead service line replacement.

The American Society of Engineers has estimated that there are 240,000 water main breaks annually due to deteriorating and poorly maintained pipes. Water is also lost through leaks and breaks in mains and services lines that go undetected and unrepaired. These pipe disruptions threaten public health, by offering the opportunity for pathogens and other contaminants to enter the drinking water distribution system. They also lead to an estimated 14-18% loss of treated drinking water. Pipe repair and replacement delivers other public health and efficiency benefits. Shoring up our underground drinking water infrastructure will guarantee more public health protection, less lost revenue for water systems and less disruption from unexpected main breaks.

While not in this subcommittee's jurisdiction, we note that increased investment in Clean Water State Revolving Funds is also critical, and will contribute to improved water quality nationwide and thus to cleaner drinking water sources.

DRINKING WATER SRF'S - PROTECTING OUR NATION'S DRINKING WATER SOURCES

Drinking Water State Revolving Funds can be used for source water protection activities, and many communities are using innovative strategies to leverage use of SRF funds and other sources to keep drinking water sources cleaner. The return on investment for protecting drinking water sources is clear. According to EPA, every dollar spent protecting a drinking water source results in savings of up to $27 on water treatment. We support proposals to allow States to use Drinking Water SRF funds to work with water systems to update source water assessments, identify vulnerabilities and prepare protection plans.

PRIORITIZE COMMUNITIES MOST IN NEED

We support additional targeted funding for disadvantaged communities as well as more flexibility in the form of grants or loan forgiveness for communities that face major health threats. EPA and states should be able to prioritize investment in communities of color and low-income communities that have been demonstrated to be most at risk and where other sources of investment, including municipal and state support, have not been forthcoming.

RESEARCH AND INNOVATION CAN IMPROVE EFFICIENCY, TRANSPARENCY AND DRINKING WATER QUALITY

We urge the Subcommittee to support increased investment in drinking water research and innovation. We lack sufficient data in a number of areas that are critical for protecting public health and for keeping Safe Drinking Water Act program up to date, including contaminant occurrence in drinking water sources, health effects from drinking water contaminants, and treatment. Innovation data and information systems could be valuable in numerous areas, including increased transparency, better public engagement and awareness, more effective oversight, more sustainable water systems, and ultimately increased public health protection.

STATE PROGRAMS NEED INCREASED SUPPORT

Most states implement the Safe Drinking Water Act in partnership with EPA. Inadequate state agency resources for Safe Drinking Water Act implementation is a chronic problem. The Association of State Drinking Water Administrators (ASDWA) estimates that the gap between needs and current funding for comprehensive state programs is $308 million. Meeting the gaps in state drinking water program will increase public health protection, lead to more effective implementation of drinking water protections, and support progress toward 21st century drinking water systems.

EPA Activities are Critical to Effective Investment

EPA programs that implement the Safe Drinking Water Act and support state programs and water system compliance are critical to the success of those state programs and to the quality of our nation's drinking water. Increased Drinking Water State Revolving Fund investment will not lead to better drinking water quality or more sustainable drinking water systems in the face of inadequate EPA staffing and funding to support oversight, enforcement, research, development of contaminant standards, support for small systems and other critical activities.

Thank you for the opportunity to provide these comments.

Mr. Shimkus. Thank you.

The chair now recognizes Mr. James Proctor, Senior Vice President and General Counsel at McWane, Inc.

You are recognized for 5 minutes.

Statement of James Proctor

Mr. Proctor. Chairman Shimkus, Ranking Member Tonko, Chairman Walden, Ranking Member Pallone, members of the subcommittee, good morning. I am Jim Proctor from McWane in Birmingham, Alabama, and I greatly appreciate the opportunity to be here this morning to testify about an issue that is so vital to our Nation's health, economy, and security.

For almost 200 years, McWane has proudly provided the building blocks for our Nation's water infrastructure, supplying the pipes, valves, fittings, and related products that transport clean water to communities and homes across the country. We employ more than 6,000 team members who work in 14 States and 9 other countries. Most of those team members are represented by the United Steelworkers and other labor organizations who we consider as partners in our efforts to improve our economy and our communities.

I am pleased that the Committee is considering efforts to modernize the Drinking Water State Revolving Fund. The Drinking Water SRF has played a key role in delivering the investment efficiently to communities throughout the Nation. However, as the Committee has recognized, it needs reform to make it more responsive to the scale of America's water infrastructure needs.

A vital component of any drinking water SRF improvement is a significant and consistent annual authorization level to spur increased capital investment. This investment will create and preserve the highways jobs that make these products and allow producers to harness the economies of scale that make American products more competitive. These impacts have a multiplier effect as they ripple through supply chains.

We also need to invest those dollars wisely. Like generations before us, we should rebuild our infrastructure with the most durable energy efficient and safe materials available. And smart technology offers many innovative solutions that can improve system management and reduce cost to cash-strapped utilities. But increased funding and better management do American workers and industry little good if their tax dollars are spent on unfairly traded foreign imports.

Like many other American manufacturers, we have made huge investments to modernize our operations to exceed the world's most rigorous environmental safety and regulatory standards. But we must compete every day against foreign state-owned or state-sub-sidized foundries that do not operate by any comparable regulatory standards and have little regard for workplace safety or the environment. This creates significant competitive disadvantages and have led to lost sales, closed plants, and lost jobs. And as the factories that once built our Nation's infrastructure disappear, communities lose the vital tax revenues and rate payers needed to operate and maintain their water systems.

Put simply, we can't continue to divorce Federal regulatory policies from procurement policies. The same Federal Government that regulates our operations and taxes our workers should use their tax dollars to purchase domestic products for the Nation's infrastructure, particularly when foreign alternatives are produced in conditions that would make members of this esteemed body cringe.

Fortunately, this problem has been mitigated recently by the application of the American Iron and Steel Buy American preference to the SRFs and WIFIA programs. Buy America has created incentives to preserve increased production capacities in the United States and to maintain work forces critical to sustaining the communities around them. I can say with pride and relief that this Buy American preference has saved at least one of our plants and preserved hundreds of jobs in the economically depressed area.

By 2008, our waterworks fittings plant in Anniston, Alabama, was the last surviving domestic manufacturer of these products. At one time, there were as many as a dozen such plants in the U.S., but all fell victim to the unfair competition I described previously. Even that lone survivor was at risk of closure during the Great Recession, operating at around 30 percent of its production capacity. But because of Buy American, that plant has increased its capacity utilization to almost 70 percent, added product offerings, and more than doubled the number of jobs. Our other plants have seen similar benefits.

But the impacts aren't limited to our operations. Because of Buy America, the primary importer of waterworks fittings has brought its production back to the United States, recently purchasing a domestic production facility and restoring hundreds of American jobs, while increasing competition in the marketplace.

In 2014, Congress codified the Buy American preference for the Clean Water SRF and WIFIA. Over that same time, it has been applied to the Drinking Water SRF through the annual appropriations process.

Congress should align the Drinking Water SRF with the Clean Water SRF, WIFIA, and other Federal infrastructure programs, like transportation, of making the provision permanent. This will not only preserve jobs, but a consistent standard will increase administrative efficiency and reduce costs since many water projects tap into multiple federal funding sources.

The reformation and reauthorization of the Safe Drinking Water Act programs with the Buy American preference are crucial to our Nation's health and prosperity. We at McWane are honored to have the opportunity to contribute to that process.

Thank you very much.

[The prepared statement of Mr. Proctor follows:]

Before the House Energy and Commerce Committee

Subcommittee on Environment and the Economy

Hearing on H.R.______, Drinking Water Improvement Act and Related Issues of Funding, Management, and Compliance Assistance under the Safe Drinking Water Act

May 19, 2017

Testimony of James M. Proctor II

McWane, Inc.

Senior Vice President and General Counsel

Chairman Shimkus, Ranking Member Tonko, and members of the subcommittee:

Thank you for the opportunity to testify about an issue vital to our nation's health, economy and security. Water is our most precious resource, one that is essential to human life and health. Access to water depends upon a reliable water infrastructure system that preserves, treats, and delivers safe drinking water to our nation's communities. For almost 200 years McWane, Inc. has proudly provided the building blocks for our nation's water infrastructure, supplying the pipe, valves, fittings and related products that transport clean water to communities and homes across the country and around the world. In the process we employ more than 6000 team members who work in 25 manufacturing facilities in 14 states and in additional operations in 9 other countries.

Despite its obvious importance, "out of sight, out of mind" best describes the nation's attitude toward water infrastructure. Potholes, train wrecks, and delayed flights are much more visible; thus, transportation needs often crowd out our attention to water as a serious infrastructure need. But the reality is that much of America's drinking water, wastewater, and stormwater infrastructure, including the more than one million miles of pipes beneath our streets, is nearing the end of its useful life and must be replaced. Many communities strain to maintain and operate their water treatment systems. According to the U.S. Census Bureau, nearly half a million U.S. households still do not have access to safe drinking water or a working toilet. As much as 25-30% of the treated water that goes into our distribution systems leaks into the ground as it flows through pipes installed as many as 150 years ago.

Those losses not only squander a vital and sometimes scarce resource; they represent a massive waste of the energy and associated capital required to treat and pump that water. As much as 19% of our nation's electricity consumption and 30% of our natural gas consumption is related to water treatment, pumping, and recovery. The energy used to treat water that leaks into the ground is simply wasted, which in turn increases energy prices for consumers and greenhouse gas emissions associated with its production.

Compounding the problem, our shifting population brings significant growth to some areas of the country requiring larger pipe networks to provide water service, while population decreases in other areas deplete budgets necessary to sustain water systems built for larger customer bases. Water is also a vital national security issue. U.S. security experts expect that within ten years, countries of strategic interest to the U.S. will face significant water challenges and more and more will come to the U.S. for expertise.

In every crisis, there is opportunity, and the water infrastructure crisis is no different. Investment in water infrastructure means more jobs: every $1 billion invested in infrastructure creates or supports 28,500 jobs, and every dollar invested in water and wastewater infrastructure adds $6.35 to the national economy. Moreover, the investment is largely self-sustaining. Studies have shown that with the increase in GDP, every dollar of water infrastructure investment generates $1.35 in tax revenue to the federal government and $.68 to state and local governments, tax revenues to help pay for the investment. Water also offers a unifying opportunity to make progress at home, while also projecting American leadership and boosting exports of U.S. solutions, products, and services abroad.

Our country has a choice: we can continue to ignore the problem, thus increasing the long-term burden for future generations, or we can do the responsible thing and take a strategic approach to carefully prioritize and invest in water infrastructure renewal that will ensure the public health, safety, security, and economic vitality of our communities.

I am pleased that the committee is considering efforts to reauthorize and modernize the Drinking Water State Revolving Fund (DWSRF) in this hearing and the draft legislation we will be discussing today. Since its inception, the DWSRF has played a key role in delivering investment efficiently to communities throughout the nation. Still, it has never been reauthorized since its creation and it needs reform to make it more responsive to both the scale of America's water infrastructure needs and the imperative that infrastructure investment be undertaken in a manner that creates and supports good American jobs, particularly manufacturing jobs.

The most important action your legislation should include is a substantial increase to the authorization level for the DWSRF. Building water infrastructure requires capacity, and companies need market and funding certainty to ensure that investments in building that capacity will not be wasted. A long-term, high level of annual authorization for the DWSRF will provide that market signal and spur increased use of the capacity that already exists and, potentially, the development of even more capacity as the market dictates. The obvious benefit of this – and one that is top of mind for all of us – is that this will create good, family-supporting manufacturing jobs. But another benefit is that as American manufacturers ramp up production, they can harness economies of scale and that makes American products more affordable and more competitive. There are several ways that this program can be tweaked and improved, but in the end there is no substitute for a strong, long-term, stable funding stream for this program.

While this funding is crucial, it does American workers and industry little good if those taxpayer dollars are spent on unfairly traded foreign imports. We must ensure that our efforts have the maximum impact on the American economy, and that the hard-earned tax dollars paid by American workers support the creation and preservation of American jobs, and protect our environment.

U.S.-made waterworks foundries conform to the world's most rigorous and effective, but also expensive environmental standards. American companies have invested significantly, at great cost, to modernize their U.S. operations to meet federal environmental and worker safety regulations. We at McWane are proud to say that our plants are among the safest and most environmentally sound in the world, but every day we must compete against foreign, state-owned or subsidized foundries and mills that regularly flout international trade laws, have no regard for worker safety, the environment, or public health and are not required to operate by standards comparable to those with which U.S. manufacturers must comply. In fact, the foreign-origin producers with whom U.S. iron and steel producers most often compete are also the most polluting. A typical foundry in China, for instance, emits more than 20 times the particulate (9.4 lbs. per ton versus 0.4 lbs. per ton) and nearly 35 times the carbon monoxide (149.4 lbs. per ton versus 4.4 lbs. per ton) than are emitted by a typical U.S. foundry. The carbon dioxide emitted from China's iron and steel industry accounts for as much carbon dioxide emissions as the rest of the global iron and steel industry. In addition to the harm to the environment, these disparities create significant cost and competitive disadvantages for American producers, that have led to lost sales, closed plants, lost tax revenues, and lost jobs. Communities across the country are in decline because the factories that once built our nation's infrastructure have disappeared, depriving them of the vital tax revenues and rate payers needed to operate and maintain their water systems and other public services.

Fortunately, this problem has been mitigated in recent years through the application of the American Iron and Steel (AIS) preference to the DWSRF, and the Clean Water State Revolving Fund (CWSRF) and Water Infrastructure Financing and Innovation program (WIFIA). AIS is critical to U.S. iron and steel producers. It has provided producers with critical incentives to preserve production capacities in the United States, make significant capital investments to improve manufacturing capabilities, and maintain workforces critical to sustaining the communities around them. I can say with pride and relief that AIS has saved at least one of our plants from closure, preserving hundreds of jobs in an economically depressed area.

By 2008, our waterworks fittings plant in Anniston, Alabama was the last surviving domestic manufacturer of those products. At one time there were as many as a dozen such plants in the United States, but all, including our other fittings plant in Texas, fell victim to the unfair foreign competition I described previously. Even that lone survivor was at risk of closure when the great recession hit, operating at around 30% of its production capacity. But with the application of AIS to the SRF's, first in ARRA and later through WRDA and the annual appropriations process, that plant has increased its capacity utilization to almost 70%, added product offerings, and, more importantly, more than doubled the number of jobs. But the benefits of AIS are not limited to our operations. Because of AIS some of the same foreign companies who drove the near destruction of the American fittings industry have now moved their production to the United States, first using existing foundries struggling for work, and more recently purchasing their own production facility. They have done this specifically in response to AIS. It is hard to conceive of a more concrete example of AIS's job-creating impact.

AIS was first enacted for both the DWSRF and the CWSRF in the Consolidated Appropriations Act, 2014. Later in 2014, the Congress enacted as part of the 2014 Water Resources Reform and Development Act permanent AIS statutes applicable to the CWSRF as well as the new Water Infrastructure Financing and Innovation program. Since that time, we have urged Congress to enact a statute to permanently apply the AIS procurement preference policy to the DWSRF just as Congress has applied the policy annually though the appropriations process to the DWSRF for Fiscal Years 2015, 2016, and 2017.

While it is to the credit of bipartisan and bicameral Appropriators that the DWSRF provision has been annually renewed, the Appropriations approach to AIS has always been a temporary means to the appropriate end, which is permanent statutory application of the preference via authorizing legislation. Moreover, many other water related programs have no domestic content requirement, which not only deprives the economy of the benefits of AIS, it also creates administrative inconsistencies and inefficiencies. The programs with no Buy America requirement include the U.S. Department of Agriculture's Rural Utilities Services' Water and Waste Disposal Program, the U.S. Department of Housing and Urban Development's Community Development Block Grant program, the U.S. Bureau of Reclamation's Rural Water Supply program, the Economic Development Administration's Public Works and Economic Development Program, and the Indian Health Services, Facilities and Environmental Health program.

Until Buy America preferences like AIS are made permanent and applied across the spectrum of water programs, the thousands of jobs that have been created and supported by this successful policy are always at risk. It is time to build on what is already a successful program, and to make AIS permanent for the DWSRF and other water programs as it is for the CWSRF, WIFIA, and most of the other non-water federal-aid infrastructure programs. As this legislation is further developed, the subcommittee and the committee should ensure that the AIS is permanently applied.

While the reauthorization of the DWSRF with a strong level of funding and a permanent application of the AIS preference are crucial to the improvement, rehabilitation, and expansion of America's water infrastructure system, they are not the only piece. It must act in concert with other federal and state programs, and new initiatives are needed to develop a comprehensive national approach to this crucial need.

In addition to the SRFs and AIS, some additional steps that can and should be taken to address our nation's water infrastructure needs include:

Establishing a Presidential Commission on Water Infrastructure Policy Coordination and Security to Evaluate and Create a Coordinated, Rational, and Efficient Water Infrastructure Policy and a Process for Administration

The federal level of responsibility for water infrastructure and quality is currently shared across approximately thirty agencies, ten departments, and several independent commissions, councils, and offices. Mobilizing and aligning U.S. government agencies, the private sector, and civil society organizations through the creation of a short-lived (one year), focused Commission on Water Infrastructure Policy Coordination and Security (the "Water Commission") can play a positive role in breaking down the silos that have been unavoidable in such a diffuse system. Further, it can create a coordinated, rational, and efficient water policy and administration and foster better collaboration and coordination across federal agencies and other stakeholders. This Water Commission should develop a national strategic plan for water investments, and report back to the Congress with a plan for the strategic direction, coordination, and oversight to domestic and international water-related activities.

Increase Public and Private Investment in America's Water Infrastructure and Create Jobs

Rebuilding our nation's water and wastewater infrastructure will require increased public and private capital investments. The American Water Works Association and EPA have estimated it will cost between $650 billion and $1 trillion over the next 25 years to maintain current levels of water service. At the same time, private investors with billions of dollars of private capital are searching for ways to invest in water infrastructure. By creating the right mix of incentives, the United States can enhance the ability of state and local service providers to raise the capital they need and encourage significantly more private investment to help modernize America's water and wastewater infrastructure, thus putting millions of Americans to work in renewing our infrastructure for the 21st century. Options that would address this funding gap include:

Remove the Volume Cap on Private Activity Bonds (PABs)

Congress should amend the Internal Revenue Code of 1986 to remove the volume cap for private activity bonds used to finance water and sewage facilities. The annual volume cap hinders the use of PABs for water and wastewater infrastructure, which are generally multi-year projects and outside the public eye. In recent years as little as 1-1.5% of all exempt facility bonds were issued to water and wastewater projects. Removing water and wastewater projects from the restrictive state volume caps will increase private capital investment in the nation's aging water infrastructure by up to $5 billion annually according to the EPA, increasing jobs, GDP, and tax revenues while solving a tremendous public need. According to the Congressional Budget Office, over ten years this policy change could infuse $50 billion in private capital investment at a cost of only $354 million in lost tax revenue.

Significantly Increase Congressional Appropriations for State Revolving Funds (SRFs), Enable Private Sector Participation in SRF Projects

Congress should authorize and appropriate funding for SRFs, which are the nation's principal federal-aid programs for clean and drinking water infrastructure, at the level of the greater of 20% of the funding in any 2017 infrastructure bill, or $10 billion annually for the Clean Water State Revolving Fund (CWSRF) and $10 billion annually for the Drinking Water State Revolving Fund (DWSRF). In addition, SRF authorizing legislation and implementing regulations and guidelines should be amended to encourage additional private investment.

Increase the Funding for the Water Infrastructure Finance and Innovation Act (WIFIA) Program

WIFIA is emerging as an extremely effective and cost-effective tool for addressing financing needs in the water sector. WIFIA is in a position to promote the use of public-private partnerships in this area by reducing the cost of private participation. Earlier this month, EPA announced that it had received interest from 43 entities for the first round of WIFIA loans. Speaking about the announcement, EPA Administrator Scott Pruitt said, "[a]s a federal-local-private partnership, this program will help expand water infrastructure systems to meet the needs of growing communities. This investment will empower states, municipalities, companies, and public-private partnerships to solve real environmental problems in our communities, like the need for clean and safe water."

In 2017, Congress should appropriate $1 billion for WIFIA. When used to provide credit enhancements, every dollar provided by WIFIA will generate $65 in additional, private capital. Thus $1 billion of funding could generate as much as $65 billion in infrastructure investment.

Eliminate or Modify Tax Rules and Regulations Related to Defeasance that Create Obstacles to Public-Private Partnerships

An estimated $100 billion in private capital is available to invest in the domestic U.S. domestic water and wastewater market, which some experts have valued at approximately $130 billion. However, current regulations discourage many municipalities from entering into cost-saving and efficiency-driven partnerships with private water companies for the operation of municipal water supply and treatment facilities. Specifically, IRS regulations impose a significant financial penalty on municipalities who sell or lease their water system to a private company if it was originally financed with tax-exempt debt. Removing tax inefficiencies for lease and sale of municipal water systems will provide greater options and opportunities for communities with failing water systems to attract more private investment and expertise to rehabilitate and restore failing water infrastructure though public-private partnerships.

Retain Tax Exemptions for Municipal Bonds

Tax-exempt municipal bonds are the primary means by which utilities and municipalities raise capital for water infrastructure projects. The market for these bonds provides an established, reliable, and efficient mechanism for public utilities to raise low cost capital. The tax-exempt feature of these bonds should be preserved in any tax reform measures adopted by Congress.

Encourage Water Utilities and Operators to Fully Account for Total Costs

In part because of a steady decline in federal funding for water infrastructure, approximately 98% of water projects are financed by local water utilities through their rate structures. However, water is arguably the world's most undervalued resource, as traditional approaches to pricing have not reflected the true cost of service. A recent survey found that only one-third of water utilities are operating under rate structures that provide adequate revenue to fully cover their costs. This undervaluation of water as a commodity creates severe constraints on the ability of utilities to finance the investment required as their infrastructure continues to age. To correct this problem, utilities must price their water based upon its true cost, while ensuring that lower-income households have reliable and affordable water service.

State and local water agencies are best able to assess how best to meet the needs of their water consumers. Where federal support is requested, however, applicants should be encouraged to conduct a study of the total costs associated with constructing, operating, and maintaining their water, wastewater and storm water systems, including long-term capital costs.

At the same time, low-income customers should be protected against significant rate increases that jeopardize their health and well-being. To create this safety net Congress should establish a support fund modeled on the Low-Income Home Energy Assistance Program called the "Low-Income Water Assistance Program" (LIWAP).

Grant Greater Flexibility to the States to Make Use of Unliquidated Obligation (ULO) Balances to Provide an Additional Source of Funding for Projects

"ULO balances" refer to unspent funds from grants provided by EPA to the states to support the financing of infrastructure improvements to drinking water systems and other important public health protection purposes. Funding that has already been allocated to the states in previous years – but simply remains unspent – should be applied to meet current-year and future-year water infrastructure needs and thus reduce the level of additional appropriations necessary.

MAKE AMERICA'S WATER INFRASTRUCTURE WORK BETTER

Congress should encourage actions that will unleash America's know-how, strengthen the technical and managerial skills of our workforce, vastly improve the efficiency and resiliency of our water systems, and promote the development, deployment, and diffusion of 21st century solutions throughout the United States and around the world.

Promote Smart Technologies and Smart Cities

Without a full set of data and actionable information about network infrastructure conditions, operators are trapped in a reactive cycle. In addition, most utilities continue to utilize monitoring techniques that provide incomplete and often stale information about critical situations such as toxicity and chemistry, a situation that can endanger the public's health and the integrity of the water systems. Wireless technology and new sensing and metering capabilities create opportunities for remote but inexpensive real-time flow and quality monitoring. According to research commissioned by utility infrastructure company, Sensus, digital water networks can save utilities up to $12.5 billion a year. Policy tools that could remove barriers to digital water adoption include:

Establish the "National Water Infrastructure Test Bed Network"

Unless utility operators have the confidence that new technologies will work, they are reluctant to adopt or deploy them. But few are willing to serve as the pilot program because of the demands on time and budget, and even those pilot programs that do proceed can take years to complete. As a result, the deployment of workable, cost-saving and efficiency-creating technologies is unnecessarily delayed.

Congress should authorize and fund the creation of a "National Water Infrastructure Test Bed Network" (TBN), to coordinate and accelerate the water industry's deployment of new technologies. It would bring together the broader water community (i.e., regulators, operators, consulting engineers, etc.), and engage them in piloting and demonstration efforts to raise confidence in innovative technologies. The TBN's process would reduce the number of pilot projects otherwise needed and would also shorten the time needed to achieve commercial acceptance.

Regulatory Reforms to Promote Adoption of Better Infrastructure Technology

Duplicative, unnecessary and/or outdated regulations present a significant barrier to addressing water infrastructure issues. Public water authorities are loath to take substantial risks in new and efficient technology procurement, because they must manage an essential public service for perpetuity and at minimum cost. Some specific examples of opportunities for EPA reform:

Reform Technology Approval Process

Both the Safe Drinking Water Act and Clean Water Act require utilities to use EPA approved protocols for monitoring and treatment. While EPA's drinking water offices have implemented an Alternative Technology Approval process that has significantly expedited the commercialization of new technologies, the wastewater program has not adopted this new, more efficient process. As a result, the deployment of new technologies in the wastewater sector has been slow. Congress should require the agency to adopt the same approval processes across its programs so that they are consistent and efficient. In addition, the agency should use existing consensus bodies (e.g., ASTM, AWWA) to the maximum extent possible to foster best management practices and standards that support the technology adoption, while minimizing agency expenditures on this mission.

Reform National Science Advisory Board

EPA's National Science Advisory Board plays an essential function in advising the agency on all manner of technical issues affecting regulatory promulgation and the water sector. This board should be comprised of representatives from across the water sector in a balanced fashion, so that the agency fully understands state-of-the-art science, industry practices, and the challenges facing water infrastructure owners and managers and their rate-payers.

Enact Legislation to Promote 21st Century Digital Water Solutions

The Internet of Things can enable water utilities to connect their physical assets with other key pieces of information, such as soil conditions, water chemistry and quality, and geospatial seismic activity to expedite repairing of infrastructure problems, reducing service disruptions, decreasing water losses, avoiding public health emergencies, and predicting other potential problems for early intervention. Expanding existing water infrastructure funding programs (SRFs, WIFIA, and private activity bonds) to include digital water projects as eligible activities will promote rapid and wide-spread adoption of digital water solutions, and in turn create high-value jobs for digital water solutions providers and utilities.

Empower Local Decision Making

Communities across the country have diverse water and wastewater infrastructure needs. They must evaluate numerous factors when considering the proper design and materials for their community and water projects. Encouraging and supporting local governance allows those closest to the problem to determine the best solution. Deference to local decision-making also saves money, as local communities can hold those in their community more accountable. Congress should encourage federal agencies to defer to local communities and their engineers of record.

Encourage Life-Cycle Costing and Pricing to Ensure Long-Term Value

To ensure that the true costs of building, maintaining, and operating our water systems are captured and funded, all federally supported projects should encourage the use of full life-cycle cost analyses when comparing bids, so that they consider the true cost of systems and materials over their entire useful lives.

Improve Systems Management

Unleash the "Blue Wave" to Build Capacity for Water and Wastewater Utilities and Other Water Resources Managers

Congress should authorize the creation of a platform for collaboration among public enterprises, in the form of a web-based portal and network – the "Blue Wave." This portal would enable urban and rural utilities of every size and service to share best practices, develop joint partnerships with public and private agencies, engage private sector expertise and technology and access private capital markets and funding. In addition, this network would provide small rural and distressed water systems with the technical capacity to comply with regulations and to undertake projects to improve or expand their services.

The United States has more than 60,000 water utilities, the majority of which are small utilities (serving less than 10,000 customers). Thousands of small utilities have difficulty in assessing, selecting, implementing, and financing necessary capital upgrades or in implementing new smart technologies that can improve service and reduce operating costs. By creating the "Blue Wave" to deliver up-front expertise and transaction support and foster public-private partnerships, the federal government would enable small utilities to carry out quickly a backlog of vitally needed capital improvements worth hundreds of billions of dollars. This would generate tens of thousands of jobs and boost U.S. businesses engaged in design, construction, maintenance, operations, and technology and scientific support.

A relatively small investment of the federal government to support the start-up costs (approximately $5 million per year for two years from the appropriated funding for the state revolving funds), to be matched 2:1 by the private sector, would fully support the development of the network. Membership in "Blue Wave" would be cost-free to agencies seeking assistance, although a fee would be assessed on successful partnerships and collaborations to cover the on-going costs of the Blue Wave Network.

"Blue Wave" is designed to be the implementation arm of many of the parallel water infrastructure proposals to small and distressed communities, as well as the dissemination arm of the policies of either the Water Commission or the White House Council on Environmental Quality. The Blue Wave Network would be a private enterprise supported by and leveraging current platforms of public and private agencies, trade associations, and sector coalitions.

Develop a Water Workforce for the 21st Century

Attracting and training the next generation of water and wastewater system operators is critically important, particularly for small and disadvantaged communities. Many water and wastewater utilities undertake the complex challenge of consistently delivering safe drinking water with a small and under- resourced staff with limited technical skills and training. Even large utilities will soon face loss of talented workers with the skills essential to the effective operation of their systems, and the introduction of new technologies will aggravate this problem because the operators of the future will need greater technological skills than are common today.

The Safe Drinking Water Act includes several set-asides related to operator certification and training for water systems from the funding authorized for the state revolving funds. Congress should buttress that authority by tasking the U.S. Department of Labor with developing a workforce development program helping American workers get the skills and credentials needed to support the operation, maintenance, and improvement of water and wastewater systems of tomorrow.

Promote Integrated Watershed Management and Planning

Frequently, agencies and local officials only manage the water bodies near urban areas rather than upstream sources. As a result, when the water arrives for treatment it often contains severe contamination from upstream activities. This not only creates a public health threat, it also dramatically increases the costs to utilities to treat the water to safe standards. More effective management at the watershed level will significantly reduce operational costs to utilities and customers by limiting the amount of local water treatment necessary to ensure good water quality.

Codify EPA's Integrated Planning Process as An Option for Local Governments to address their Wastewater and Storm water Management Needs

For years, communities have been told to achieve Clean Water Act mandates without any consideration of whether those requirements are feasible, affordable, or provide a significant environmental benefit. This problem is especially acute for separate sanitary sewer systems. To partially address this issue, H.R. 6182, the Water Quality Improvement Act, introduced in the last Congress, would allow local governments to prioritize and focus on their wastewater and storm water management needs with the greatest public health and environmental benefits. This legislation would codify EPA's Integrated Planning process as an alternative to facing costly consent decrees while establishing economic affordability criteria for the EPA to assess the financial capability of communities to implement control measures.

Amend the Clean Water Act to Improve Federal Agency Coordination and Advance Integrated Water Resources Management (IWRM)

Congress should codify integrated water resource management (IWRM) principles into the Clean Water Act (CWA) to promote greater coordination among federal agencies such as the Army Corp of Engineers, the Department of Agriculture, the Department of Interior, and EPA, which often pursue separate and sometimes conflicting agendas for water resource management. Amending the CWA to promote greater coordination through the Water Commission and IWRM would foster more coherent water management among relevant federal, state, and local authorities, thereby optimizing the use of water for agricultural, urban, and ecosystem needs.

Reform the Agricultural Act of 2014 to Reduce Agricultural Runoff in Watersheds

Congress should reform the Agricultural Act ("Farm Bill") to upgrade water quality standards at the watershed level for runoff originating from agricultural areas that affect downstream water users. Existing laws do not adequately protect sources of drinking water from contamination by agricultural runoff, which includes pesticides, herbicides, or animal wastes, thus imposing greater treatment costs on downstream water utilities. Farmers should be encouraged to use existing incentive mechanisms to deploy best management practices to meet the standards.

Promote Water and Wastewater Regulatory Reform

A key reform measure would be to require the consideration of opportunity costs in assessing the economic toll of any new, proposed regulations. The true cost of compliance with a regulation not only includes the direct costs incurred, but also the loss of other opportunities because of the divergence of scarce resources. For example, burdensome recordkeeping or treatment requirements that do not materially improve public health might require the expenditure of thousands of dollars that could have been spent on infrastructure repairs or the addition of smart technologies that would have reduced water loss or otherwise improved efficiency. The benefits lost from those alternatives should be included in the calculation of the true cost of a new regulation, so that policy makers have a more accurate understanding of the consequences of regulations.

MAKE AMERICA'S WATER INFRASTRUCTURE SAFER AND MORE SECURE

Americans must have confidence in the safety and reliability of their water supplies. Likewise, by promoting improved water management and greater cooperation on shared waters within the United States and internationally, we minimize the potential for disruptive conflicts over water, thereby strengthening America's security and economic interests at home and abroad.

Service lines, including those made of lead, can be a major source of toxins entering households across the country. Service line ownership is typically split between the homeowner and the water system. An array of solutions exists for dealing with toxins in water, ranging from better treatment and system management to in-home filtration to full replacement of outdated service lines. Full replacement is very expensive, and replacing all lead service lines across the country could cost a total of approximately $30 billion. Many homeowners have limited discretionary funds to fund replacement of their lead service lines. Moreover, lower-income families tend to live in older housing units that are more likely to contain lead service lines, so the potential burden for a replacement program might disproportionally affect the poor.

To address these concerns Congress or the relevant agencies should consider:

- Creating tax incentives for replacement of service lines or installation of in-home filtration systems to remove lead and other toxics from drinking water;
- Expanding EPA and HUD real estate disclosure requirements to include whether a lead service line is present, and including lead in drinking water in the EPA and HUD's Lead-Safe Housing Rules;
- Adapting federal assistance programs such as the Federal Housing Administration (FHA) to make lead service line replacement an eligible activity for rehab mortgage insurance under Section 203(k) or as a Title I insured loan for property improvements; and,
- Increasing funding allocated to existing grant programs such as HUD's Community Development Block Grant program for purposes of lead service line remediation by low-income households, either through filtration or replacement.

Conclusion

These are only a few of the issues and solutions that merit discussion. The key takeaway, however, is that the scope and scale of America's water infrastructure needs demand a massive, coordinated, forward-thinking, and creative response. Water infrastructure is not a partisan or even a bi-partisan issue. It is and must be a non-partisan issue. With that cooperative spirit in mind, reform and reauthorization of Safe Drinking Water Act programs like the Drinking Water SRF are crucial to that effort, and we at McWane are glad to have the opportunity to contribute to that process.

Thank you for your time and consideration.

Mr. Shimkus. I thank you all for your testimony.

We will now move into the question-and-answer portion of the hearing. I will begin the questioning and recognize myself for 5 minutes.

And, of course, I will go to Mr. Fletcher first. Is it challenging for a small community to go through application processes for government assistance?

Mr. Fletcher. Very much so, Congressman.

Mr. Shimkus. What would you recommend a process of streamlining or the challenges? What could we do to make it easier?

Mr. Fletcher. Well, I believe that if we have assistance, circuit rider program, something similar to that, for each state, that the circuit riders would have the knowledge to go to these small systems and help them through the process with the SRF application.

Mr. Shimkus. Mr. Vause, your testimony calls for streamlining the SRF application process. What does that include for you?

Mr. Vause. Mr. Chair, we do support efforts to reduce the burden on regulation and the application process itself. We think that the EPA can do, among its regions, developing best practices that can be applied to all of the regions there to streamline the application processes themselves.

We believe, secondarily, that the ability to do the applications themselves rely on certain forms and certain procedures that the agency should streamline. Those procedures themselves go to the issue of the Buy America provisions, they go to the issues of tracking minority, disadvantaged, and women business enterprise activities related to SRF projects. So those are two areas that we would like to see where there is streamlining done. Thank you.

Mr. SHIMKUS. And if anyone else on the panel would like to comment on the possibility of streamlining the application process for SRF?

Oh, Ms. Daniels.

Ms. DANIELS. Yes. So if I could just add. So we have heard from applicants that they much prefer the RUS program because it is much more streamlined. And it seems that it can give the applicant upfront information sooner about what they might be eligible for, what rates they might be looking at, and it helps them then move forward from there and really design the project that fits sort of their understanding of funding.

So if our program could figure out a way maybe to do a letter of intent where you get the financial information up front, because that is generally what is used to determine rates and moneys available, that would give folks some upfront information then to move forward and finish the complete application.

Mr. Shimkus. Yes. What is the burden? You mentioned burden.

Ms. Daniels. So the burden for completing the application?

Mr. Shimkus. Right.

Ms. Daniels. Well, I mean, it is substantial for small systems.

In some cases, they are just not capable of completing it. So one of the assistance programs that I mentioned before, professional engineering services program, we do provide assistance. So if a community really needs help completing the application, we will work with them to do that.

Mr. Shimkus. And I agree, being from rural America, I think the RUS ability for rural water co-ops and stuff have been very, very helpful. And I haven't heard the same concerns that I had with the SRF.

Going back to you, Ms. Daniels, are there other reasonable steps that can be taken to simplify the SRF application process or paper-work? Anything else you can think of?

Ms. Daniels. I think if we can come up with sort of an upfront screening process, so an upfront letter of intent, I think that gives folks a better sense.

So in Pennsylvania, before they can come in for an application, they already have to have the project designed, they have to have all of the permits in place. There is a lot of expense that goes into getting to that point, and we don't even know yet, then, what they might qualify for or what rates they might be looking at.

Mr. Shimkus. So let me finish up with you. We have heard a fair amount of testimony on disadvantaged communities. Are you comfortable with the flexibility that the Safe Drinking Water Act allows regarding the amount you can spend and how much debt you can forgive?

Ms. Daniels. We really are. I think keeping the language of ''up to'' gives states the flexibility. So in a given year, if we have lots of projects that meet that criteria, we are able to fund those. But in other years where we don't, it means we don't necessarily have to set that funding aside. We can use that for other worthwhile projects.

Mr. Shimkus. Thank you very much.

I would yield back my time, and now recognize Mr. Tonko for 5 minutes.

Mr. Tonko. Thank you.

Many of the organizations represented today testified earlier this year. At that hearing, everyone agreed that more funding is necessary for the Drinking Water SRF.

The SRF was initially authorized at $1 billion in 1996 and, frankly, I don't think that level of 20 years ago would meet our Nation's needs, especially since we have seen the need grow significantly during this time period.

So my question to everyone on the panel is, do you support sustained increased funding for the SRF relative to historic levels?

Mr. Kropelnicki?

Mr. Kropelnicki. Yes, we do.

Mr. Tonko. Mr. Potter?

Mr. Potter. Sir, I would like to address the fact that the drinking water industry is a jobs program waiting to happen. We can put a lot of people to work in a hurry. So the level of funding that Congress would appropriate really can't be enough. We can put people to work. We can renew infrastructure. We can keep the dollars in the United States. We have used McWane pipe. It is a good pipe. Everything about the whole program is good for us. Fund us; we will put people to work.

Mr. Tonko. Thank you.

And can we continue, Mr. Fletcher, just across the board?

Mr. Fletcher. Any increased funding for small communities would be greatly appreciated.

Mr. Tonko. Thank you.

Ms. Daniels?

Ms. Daniels. So ASDWA supports funding of about a billion. Now, that isn't quite the same as maybe the double or the triple numbers that you are hearing from other folks. One of the reasons is that we have to understand that state staffing levels are what they are right now based on sort of the historical funding. States would have a difficult time quickly staffing up to be able to move a two or three times the amount of funding. I think what states may need is more moderate increases over a longer period of time and maybe some predictability that those funding levels will continue. That is what states need to really be sort of confident that they can increase staffing levels to be able to move those moneys.

Mr. Tonko. Right. And I believe AQUA reflects that in its language.

Mr. Vause?

Mr. Vause. Mr. Tonko, yes. As we had indicated in our testimony, the doubling of SRFs, and we believe a sustained effort is necessary both for the SRFs and the WIFIA program.

We do recognize, though, that states do have a match to the SRFs. So along with the increased funding at the Federal level is a requirement that the states have to match as well.

Mr. Tonko. Thank you.

Ms. Thorp?

Ms. Thorp. Yes, thank you, Congressman. Yes. As I mentioned, we support significant increases in the state revolving funds as well as in the Public Water System Supervision grants. We recognize there are complications and that it is not the only solution to our Nation's drinking water challenges, but it is certainly a much needed piece of the puzzle.

Mr. Tonko. Thank you.

Mr. Proctor?

Mr. Proctor. Absolutely. As has been noted previously, there is an estimated trillion dollar need to rehabilitate our country's water infrastructure. The unfortunate thing, though, is that highways, airports, other things like that get more attention, but the need is just as critical for water. If there is a pothole in a highway, I am sure you all get a phone call from a constituent, but with water, even though 20 percent of our water is leaking into the ground today, which is massive waste of a precious resource as well as the energy associated with it, it is out of sight, out of mind. But we can live without roads; we can't live without water.

Mr. Tonko. Thank you.

There are disadvantage systems that need extra assistance, and this discussion draft has some good ideas, but I believe there are additional things we can do to support them.

Mr. Potter, can you expand why it is important to expand the definition of disadvantage community?

Mr. Potter. Yes, sir. Fundamentally we are a large system, so we have 190,000 water accounts. We have areas at Metro Water Services that are relatively affluent. We have areas that are economically disadvantaged. If we do not expand the definition, then we wouldn't have the ability to have the additional subsidization available through the Drinking Water SRF.

It provides us another tool to fund a project specifically in a disadvantaged area that we would not have if the definition wasn't expanded, so we would request that it be done so.

Mr. Tonko. Thank you. And an asset management, the benefits of that management, of asset management are being more widely accepted, and I do understand the concerns about being overly prescriptive, but also believe that more can be done to encourage utilities to implement plans.

To Mr. Vause and Mr. Potter, do you see a benefit to having systems finance projects that focus on the long-term sustainability of their systems?

Start with you, Mr. Vause.

Mr. Vause. Mr. Tonko, yes, and we do believe in the encouragement of every utility doing a project of that nature to consider the life cycle costs associated with that and to factor that into the decisionmaking on what is the right solution for that particular project issue at hand.

Mr. Tonko. And Mr. Potter?

Mr. Potter. Yes, sir. Asset management is a good thing. Recognizing that some utilities will have staffing that is more available than a small system. A good example is, is this a pump? If you take a brand new pump out of the box, and you install it, and you do vibration analysis and lubricational analysis over the life cycle of the pump, it is going to last longer. And that is a better use of O&M funding.

If you don't do that, and that means you don't have asset management program, it is going to cost more. And if it costs more, those dollars will not be available for capital investment.

So overall it is a good idea. We recognize that some utilities will have higher capabilities than others, but overall asset management works.

Mr. Tonko. Thank you. Thank you. And I yield back.

Mr. Shimkus. The gentleman's time is expired. The chair now recognizes the Chairman of the Full Committee Mr. Walden for 5 minutes.

Mr. Walden. Thank you, Mr. Chairman.

Mr. Vause, one of the proposed SRF enhancements that you discussed in your testimony was added flexibility and repayment terms for the SRF loans. Why is added flexibility for repayment terms needed, and do you support the provision in the discussion draft that extends loan repayment schedules for disadvantaged communities from 30 years to 40 years?

Mr. Vause. Mr. Walden, we do support the issue of extending the terms to disadvantaged communities, and essentially it is an issue of this, when you think about when you take out a loan for a home for other things, those are long-lived assets, and to be able to extend the terms out to not exceed the useful lives of the a ssets that are being funded through the SRF and so forth, that is an appropriate way to help communities who need to extend out the terms and so forth to be able to afford the loan.

Mr. Walden. All right. And today's discussion draft removes Federal reporting requirements on Federal funding if state or local requirements are equivalent to the Federal requirements.

From your perspective, Mr. Vause, what effect would this provision have, and would it be as beneficial as some of us think it would be, and do you support it?

Mr. Vause. Mr. Walden, we do support that concept, and it does help and facilitate the ability of the loan recipient to be able to ease the administrative burden of a project of this nature.

When utilities go through, being able to show that an equal or more stringent requirement exists, at the state level, makes it much easier to facilitate the use of the loans in the administration of a project that is funded and financed that way.

Mr. Walden. And is there something we should do in terms of prioritization or should we stay out of that, and by ''that'' I mean when we identify in the country a problem, let's say lead in the pipes or arsenic in the water or something, should we be thinking about a way, or maybe it is already there, to target a support to communities to deal specifically with those issues as opposed to just a leaky water system or something of that nature?

Mr. Vause. Mr. Walden, every state that acts as the primacy agency for SRF funds has their own set of criteria that they use to prioritize projects, and typically those prioritizations involve things that are of critical public health need, and, therefore, most of the monies that our experience is, is projects go to those that have the highest priority to protect public health.

Mr. Walden. OK. Then sort of the several billion dollar question that is before all of us: How do we pay for this? I know at the local level in my water bill I pay for it. The Federal level we tend to just throw a number on a piece of paper and then go borrow it or find it or something.

Are there any of these authorized programs out there that you would tell us really aren't working and we should move money from them to this? Any ideas on how we should pay for this from the Federal level, other than giving our kids and grandkids the due bill later in their life?

Mr. Vause. Mr. Walden, I think a short answer to that is is the newly created WIFIA program is a great example of where the burden on the Federal Treasury is de minimis. In that situation it is a loan program.

Mr. Walden. Right.

Mr. Vause. And therefore, those who are in receipt of WIFIA loans really are paying back to the Federal Treasury and the effect is very, very minor.

Mr. Walden. OK. Anybody else on the panel want to tackle the funding issue, other than being recipients of the funding.

Mr. Shimkus. Would the chairman yield?

Mr. Walden. Sure, of course.

Mr. Shimkus. Under the WIFIA, which has been part of the discussion too, it is my understanding for small communities the requirements are so large that they can't apply. In fact, no loans have been made out of the WIFIA program yet.

Am I correct or someone tell me about what they have done with the WIFIA. Mr. Potter?

Mr. Potter. Sir, we think WIFIA is in addition to SRF. We don't think they are mutually exclusive. We think they are complementary, and we think they should both have equal funding attention.

Mr. Walden. But to his point, and Mr. Vause, I represent eastern Oregon, it is not as big as Alaska, but we have got a lot of these little tiny communities.

Mr. Shimkus. But you are a broadcaster in Alaska.

Mr. Walden. That is true. The Mighty Ninety KFRB Fairbanks. But the point is they don't have a huge water department, it is the mayor or somebody. They have got a public works, but what we want to do is how do we streamline this and put the money in the pipe and the ground and the water system and not in the paper-work and the reporting and all of that? Isn't that what we are trying to get to here?

Mr. Vause. Mr. Walden, with respect to the WIFIA program, for example, small communities under the size of 25,000, the project size that is eligible is a $5 million project. States also can apply for WIFIA loans, and they can bundle projects together from small communities to help facilitate that in that program.

The ability of the small communities to administer an SRF program, to that question, I think the ideas that we have previously talked about of streamlining some of the paperwork exercising, having best practices used, but more importantly, the idea of being able to demonstrate the ability to use state regulations to avoid the issues of the cross-cutting requirements at the Federal level are all things that really help try to streamline that effort.

Mr. Walden. Thank you, Mr. Chairman.

Mr. Shimkus. The Chairman's time is expired. The chair now recognizes Ranking Member, Mr. Pallone, for 5 minutes.

Mr. Pallone. Thank you, Mr. Chairman. We have seen numerous serious problems in the Safe Drinking Water Act that should be addressed in any legislation this committee passes to amend the Safe Drinking Water Act. The biggest challenge is clearly the lack of funds, but I want to quickly touch on a few others. And my questions are of Ms. Thorp.

Does the discussion draft that is before us fix the weaknesses in the standard setting process under the Safe Drinking Water Act?

Ms. Thorp. Thank you, Congressman Pallone. The discussion draft, as I read it, didn't address any of the contaminant regulation, national primary drinking water regulation setting process at all, so it didn't go into that topic.

Mr. Pallone. All right. And the source water protection provisions in the statute have proven ineffective, and that is why my bill would create an entirely new program to ensure source water protection.

Does the discussion draft before us do enough to ensure source water protection in your opinion?

Ms. Thorp. Congressman Pallone, if I recall, the discussion draft did allow for some set-asides in Drinking Water State Revolving Fund monies to do source water protection plans and to update those systems and states. So we think that is a good idea.

We do think there is some creativity and some innovation that needs to be applied as we look at the future of the Safe Drinking Water Act, which really as currently written, doesn't do much to protect source water or to reinforce our other environmental and public health protection statutes and regulations. Some interesting work could be done on that in the future.

Mr. Pallone. All right. Thank you.

Now, our Democratic proposals also address threats to source water, including oil and gas development and climate change. Does the discussion draft before us today address those threats?

Ms. Thorp. Still to me, Congressman?

Mr. Pallone. Yes, these are all to you.

Ms. Thorp. Thank you, Congressman Pallone. I did not see anything on oil and gas activities and other sector threats to drinking water sources or on climate change and resilience.

Mr. Pallone. All right. One of the concerns we hear about most on drinking water is lead contamination, particularly concerns about lead service lines and lead in school drinking water. Will this discussion draft get lead out of our homes and schools or do we need to do more?

Ms. Thorp. I don't think the discussion draft addressed lead in schools or lead in water, and specifically, although as I mention in my testimony, increased authorizations and appropriations can help us with some aspects of the lead service line problems, for example.

Mr. Pallone. All right. And then we also hear a lot of concerns about the need to restructure water systems to ensure the technical, financial, and managerial capacity to deliver safe water.

Does the discussion draft need to be strengthened to effectively address the restructuring and consolidation in your opinion?

Ms. Thorp. Well, I think some detail could be added. I think the discussion draft noted that this is one use of State Revolving Fund funds. So I think some of the detail we have seen in the bill that you, Congressman, introduced and in other places to support appropriate restructuring and consolidation would be helpful.

Mr. Pallone. All right. Obviously, it is my opinion that this discussion draft needs a lot of work if it is going to actually address the problems we see in the Safe Drinking Water Act, so my hope is that my Republican colleagues will work with us as we move forward on some of the issues that I mentioned.

I want to yield the rest of my time, though, to Mr. McNerney.

Mr. McNerney. Well, I thank the ranking member of the full committee for yielding. I am going to read a statement and I want to know if all the panel members agree with a yes or disagree with a no: "The draft mostly continues with the status quo, which is necessary but not sufficient to meet our Nation's drinking water needs."

Mr. Kropelnicki?

Mr. Kropelnicki. I would agree with that, yes.

Mr. McNerney. Mr. Potter?

Mr. Potter. Yes, sir, I would agree with that statement.

Mr. Fletcher. Yes.

Ms. Daniels. Yes.

Mr. Vause. Yes.

Ms. Thorp. Yes.

Mr. Proctor. Yes, sir.

Mr. McNerney. Well, everybody said yes. I was going to take as just the ones that said yes, name one thing briefly that you think would most improve the legislation? Starting briefly. Go ahead.

Mr. Kropelnicki. Requiring that any funds being expedited are used, be used economically, efficiently, that asset management and full life cycle pricing and full cost in the true value of water is reflected in the rates being charged to customers.

Mr. McNerney. Mr. Potter?

Mr. Potter. Yes, sir. I would support enhancement in asset management program requirements and codifying the amounts in the SRF funding levels, and strengthening the WIFIA authorizations.

Mr. McNerney. Mr. Fletcher, briefly now?

Mr. Fletcher. Technical assistance would be very important.

Mr. McNerney. Very good.

Mr. Fletcher And that for small systems in rural communities.

Mr. McNerney. [continuing]. Ms. Daniels?

Ms. Daniels. I would actually support being able to shift some of the work for source water protection plans to the SRF because that would free up set-aside funds for more technical assistance and other things within that program.

Mr. McNerney. Thank you. Mr. Vause?

Mr. Vause. Yes. EPA has stated that various states have unobligated or unspent balances in their Drinking Water SRF accounts, and when those dollars are not in circulation they are not being used to improve drinking water infrastructure.

So in combination with increased SRF funding, we, AWWA, would urge Congress to use all the necessary tools to help state primacy agencies put those unexpended funds to use in drinking water infrastructure.

Mr. McNerney. Ms. Thorp? Quickly, please.

Ms. Thorp. To increase authorization, I think creative use of technical assistance and state programs to move toward having the most 21st century modern drinking water systems we can nationwide.

Mr. McNerney. Yes.

Mr. Proctor. In addition to the domestic preference and consistent levels of funding I mentioned in my earlier remarks——

Mr. McNerney. Quickly, please.

Mr. Proctor [continuing]. Additional things that would improve, the adoption of smart technology would go a long way.

Mr. McNerney. Thank you, chairman.

Mr. Shimkus. The gentleman's time is expired. The chair now recognizes the gentleman from Texas for 5 minutes.

Mr. Barton. Thank you, Mr. Chairman. And I am not going to take 5 minutes.

We appear to be on the verge of having a bill that most people agree with on both sides of the aisle. I don't hear a lot of negativity. I guess my only question would be, this section A, it says adds a new provision that if the Federal reporting requirements on Federal funding are pretty much the same as local requirements that you don't have to make the Federal report.

Do you all agree with that? That sounds like a good deal to me.

Mr. Kropelnicki. Yes.

Mr. Barton. Nobody has heartburn over there?

Ms. Daniels. No.

Mr. Barton. With that, Mr. Chairman, I am going to yield the rest of my time to Mr. Murphy of Pennsylvania.

Mr. Murphy. I thank the gentleman. Mr. Vause, let me start off with you.

In your testimony you argued the present Buy America requirements to the SRF are unrealistic and that the conditions for granting a waiver should be loosened to make it easier to buy nonAmerican products, am I correct?

Mr. Vause. We supported modifying the language.

Mr. Murphy. Just am I correct or not, to make it easier to buy nonAmerican, is that a yes or a no?

Mr. Vause. I am sorry, could you repeat the question?

Mr. Murphy. So you said in your testimony, you argued the present buy American requirements are unrealistic and that the conditions for granting a waiver to this should be loosened to make it easier to buy non- American products. Did I understand that correctly?

Mr. Vause. Yes.

Mr. Murphy. OK. So are you willing to forego U.S. taxpayer dollars for your water projects in order to buy your steel from wherever you want?

Mr. Vause. No.

Mr. Murphy. Well, then what percent of funding from the Federal Government should you have cut in order to allow you to support the economy of China instead of the United States?

Mr. Vause. That is not our intent, sir.

Mr. Murphy. Well, but if you are not buying American steel but you are using American taxpayer's money to buy products from other countries, that is how it works out. So intention or not, that is the outcome.

So, Mr. Proctor, in your testimony you discussed the benefits to McWane and the broader domestic steel industry of the American Iron and Steel Institute preference for Drinking Water State Revolving Fund. What impact would Congress enacting a statute to permanently apply this procurement preference policy to the DWSRF have on industry, domestic manufacturing, and jobs?

Mr. Proctor. I think it would accelerate the repatriation of jobs back here to the U.S. A permanent provision would give industry the signal that it is worth investing in the new capital and the new capacity here in the United States, and we would see exactly what has already happened in the fittings business where jobs that went to China are coming back to the United States, and that would increase competition, as well as increase jobs and economic benefits.

Mr. Murphy. So you speak of the lost opportunities of the domestic industries, as well as the administrative inconsistencies and inefficiencies that this generates. Can you explain what you mean by that?

Mr. Proctor. Well, it just seems inconsistent that on the one hand you are taking tax dollars from American workers and then using those tax dollars to fund the purchase of materials and in the process taking away their livelihoods, number one.

Number two, the agency that is charged with the administration of the SRF is the Environmental Protection Agency. When they impose regulations on American manufacturers that make them uncompetitive so that people go to China, India and other places to buy their products, they are having the perverse effect of sending those manufacturing jobs to place, not only eliminating jobs here in the U.S., but sending them to places that have no regard for the environment.

Mr. Murphy. Like state-owned governments who also subsidize it and without the environment—so what happens is, so you may have an American steel worker paying U.S. taxes. Those taxes then go to help subsidize water projects to the community, which then because of the onerous regulations of the United States make other countries' steel cheaper, and those communities then buy other countries' steel, which further puts that steel worker out of a job, do I follow that correctly?

Mr. Proctor. That is exactly right. That is exactly right. And you are making the environment worse in the process. Something around 25 percent of the particulate matter that falls on California comes from China.

Mr. Murphy. So all the work we do in environmental improvements are just very small and overridden, I understand, by what China does in a short period of time?

Mr. Proctor. That is correct.

Mr. Murphy. Right.

Mr. Proctor. China produces more carbon dioxide and greenhouse gasses than all the other iron and steel manufacturing companies in the world combined.

Mr. Murphy. Thank you.

Ms. Daniels, real quickly, how big of a national problem is the undiscovered water systems containing pathogens like in Cydectin?

Ms. Daniels. It is really hard to quantify that. Every year it seems we find one or two undiscovered water systems mainly in our rural areas.

When you are driving past a community it is hard to see, are they on private wells or a connected community water system?

So often we find out about them because we get folks calling complaining about water quality, and that sort of leads us to the investigation.

Mr. Murphy. Thank you. I yield back.

Mr. Shimkus. The gentleman yields back the time. The chair now recognizes the gentleman from California, Mr. Peters, for 5 minutes.

Mr. Peters. Thank you, Mr. Chairman, and thanks for having this hearing. It comes at an important time when we obviously heard issues like Flint, we have got a 5-year drought ending in California, and it is a good time to talk about sustainability and resiliency, and we see reports that water prices would have to increase by 41 percent in the next 5 years to cover the costs of replacing infrastructure.

A New York Times op-ed by Charles Fishman said "Water is Broken. Data Can Fix It." And it claims that more than any single step, modernizing water data would unleash an era of water innovation like anything in the century. So I wanted to explore that with some of you who mentioned that.

Ms. Thorp, you said that in your testimony that invasion data and information systems could increase transparency, enhance public engagement and awareness, provide more effective oversight and ultimately lead to increased public health protection.

Can you tell me kind of what are the primary drivers for the lack of data and, you know, what are the steps we might take to employ data to be doing something beyond what we all agree we are doing today but we need to do?

Ms. Thorp. Thank you, Congressman Peters.

I think it is not a lack of data necessarily. It may be a lack of ability to compile the data and then make it usable to not only regulators but to folks in the drinking water sector in the public interest and public health communities.

There are some interesting recommendations on that sometime late last year the President's Council of Science Advisors did an interesting report on drinking water data and urged folks to take a look at it. I do think some of the authorizations we have talked about today for state programs, as well as SRFs and EPA itself could lead to progress.

Mr. Peters. I guess I am looking for more specifics on the steps we should be taking.

Sometimes I find that if you leave it up to states to make these decisions, some of them will make more progress than others if they are not given the kind of technical assistance that we might be able to provide here.

Ms. Thorp. Well, one simple step would be improving the technology we use both at EPA and in states for making it possible for drinking water consumers to understand monitoring results in their water systems, not just lead but others. That sort of thing.

Mr. Peters. Mr. Potter, maybe you had some ideas about this, as well. Is it feasible to put water quality data online in real time, would that increase transparency?

Mr. Potter. Yes sir it is. Was that directed to me?

Mr. Peters. I am sorry, I was looking at Proctor, but I am sorry,

Mr. Potter, yes.

Mr. Potter. Yes sir it is. We have real time water quality data that we do and can put on the web.

Mr. Peters. Is there something in this bill we should be doing to encourage that?

Mr. Potter. I think encouragement of that in the asset management realm would be a perfect idea. Another example would be use of automatic metering to measure use at the tap and compare that to production. That would be a great asset management tool to identify where your leaks are.

So that is lots of room for additional technology to be used in our industry.

Mr. Peters. Is that being successfully employed in particular places?

Mr. Potter. We are exploring that presently.

Mr. Peters. OK. But you are exploring whether it is being employed or how it could be employed?

Mr. Potter. How it can be used once it is deployed. We are transitioning to that technology right now.

Mr. Peters. Mr. Vause, maybe you could tell us, we received a D on our drinking water infrastructure, and you have talked about whether this bill appropriately addresses the water infrastructure needs.

What funding levels would you recommend adding into each bracket, and briefly why would you do that?

Mr. Vause. Mr. Peters, we talked earlier about the fact that we would recommend appropriations at the full authorization level for WIFIA at $45 million in fiscal year 2018, a doubling of the SRF's water and wastewater from their current fiscal year 2017 levels for fiscal year 2018.

To the issue of the data and the information, if that is part of this question, as well, I concur with what was said using it for asset management but also from security and preparedness, having on time real line data on water system quality I think is a very, very vital thing, and I think the PWSS programs and supporting the states in their efforts at not less than the current funding levels are really important to go forward.

Mr. Peters. And just along the lines of Mr. McNerney's question I think we have something here that we can find wide agreement on, but I think we can do more, and I hope we take the opportunity to improve off of the standard things we have been doing for a long time, and I appreciate all the witnesses for being here today. I yield back.

Mr. McKinley [presiding]. The gentleman's time is expired. I recognize myself for 5 minutes.

To the group, maybe it goes to you, Mr. Proctor, about energy efficiency. Tonko out of New York and Welch out of Vermont, we have worked together on trying to find ways of efficiency, and one of the things that I am concerned about is from this in the water system one of two engineers in Congress, and one of the things we are talking about is always how do we improve efficiency?

And I think a smart grid system could be very interesting with our meters, and I think you were alluding to that perhaps in your testimony because if we have 240,000 breaks during a year, and we lose anywhere from 20 to plus percent of our water, that is not efficient. The electricity is lost in motors and generating pumps to move that and the water we are moving and the chemicals all the process, so the efficiency, I know that Europe is investing about $8 billion in the next 3 years in a smart systems smart metering system.

Do you see that as being part of the solution of how we can be more prudent in our water programs?

Mr. Proctor. Absolutely. And I would like to make two points about that. One is the smart technology that is emerging right now does create the opportunity to monitor as well as meter water that is flowing through our distribution systems.

So you can detect leaks, and when you can detect the leaks you know exactly where it is so you don't spend a lot of time looking around trying to find it so you can repair it.

Mr. McKinley. If Europe is so much out in front with $8 billion, do you know what kind of numbers we are putting into this, into the research, into a smart meter?

Mr. Proctor. I don't know the answer to that.

Mr. McKinley. If you can get back to me on that.

The other thing I wanted to talk about maybe to you, Mr. Vause, is rural water. I come from West Virginia. We have a lot of areas that are really hurting for water, and I am thinking in Alaska you have got a similar situation.

And we know around the world there are some deficiencies with that people can't get access to water. And there is a program that is being developed in West Virginia at Ohio Valley University with Katharos, it is a group out of Denver in consolidation or in coordination with the Ohio Valley University to develop a mobile water treatment facility.

And they have been able to get it now to the point that they can produce water now at $0.27 per person per day. That is pretty competitive with it. So I am wondering whether or not that is something that we should—first, are you first are you aware of the Katharos Catharis program?

Mr. Vause. I am not aware of that particular program myself, but at our state, in Alaska for example, there are several ways that we are researching in partnership with the EPA ways to improve water supply to many rural areas of our state, and those include using innovation and trying to provide recycling and reuse technologies, so that for the limited supplies that are available, that there are ways in which we can improve at a household level the ability to have——

Mr. McKinley. I know their program is what they are trying to develop there, is also been using solar panels, so they can go to areas without electricity and still be able to process water for families in that immediate area.

I think it also has opportunity for us where we have some serious leaks where people can't get water that a mobile unit could come in and be able to provide them water service during the interim period of time.

I am very optimistic that these mobile units could be very helpful to us, so I thank you on that. Could you grab that? This is an example of, when I say a water problem, I have designed thousands of miles of water systems, and this is one in rural West Virginia, a good 1-inch waterline that probably has about 80 percent of it occluded that they can't pass water through.

This is what we see all across America. That is why this urgency of getting something done so that these families can have dependable clean water, and this is certainly unable to provide that.

So I thank you for that and I yield the balance of my time. Who do we have next?

OK. Mr. Green, you are recognized for 5 minutes.

Mr. Green. Thank you, Mr. Chairman. I want to thank our chairman and ranking member for holding hearings today.

Water challenges are all over the country and where I am from in Texas I have a very urban district. It is mostly incorporated by either the City of Houston or smaller cities that provide water, but we have some areas that are urban areas outside the city limits and none of the cities will annex it because of the low property values. They just can't afford to come in and put new waterlines or streets or anything else.

So what I was going to see if is in these unincorporated communities that are very urban, and I am sure rural areas have the same problem with low property values. In Texas we created decades ago water districts that are actually local levels of government for water and sewer and other things if they would like.

But, again, you can't even create that if you have low value for your property because you can't sell bonds if you can't afford to pay them off.

Is there a Federal program for these areas similar to what rural water authorities would be to help get water and sewer because, again, these are very urban areas, but our traditional sources of water and sewer are not there, so what they have is water wells and septic tanks that are, again, in urban areas not designed to have that much usage, I guess.

Is there a Federal program that would help that? Our county commissioners have helped with what they can but, again, they don't have the budget oftentimes to except to provide just a little bit of money, so that we have a partner but we would need Federal funding to do it in a low wealth area.

Anybody have any? Yes, sir.

Mr. Fletcher. Rural development has their water loan and grant program, and in Illinois, in my system itself, was unserved back in the late eighties. And we got a group of people together that tried to form this water system. And they went and talked to people, and people put deposits in of $20. It cost them $150 to get the meter once we had funding, but we went to a Farmers Home Administration and got our first loan and grant was $2-and-a-half million.

And we served those people. And we have continued to do that through this program. And I can only assume that there could be somebody in that area that would take the bull by the horns and try to do the same thing there.

Mr. Green. Mr. Proctor, can you tell us a little about the role your company plays in drinking water infrastructure projects?

Mr. Proctor. Yes, sir. We manufacture the basic building blocks for the Nation's water infrastructure. We make pipe, valves, fittings, fire hydrants, and all those related projects.

Mr. Green. OK. Coming from Houston, and I have a whole bunch of chemical plants that make PVC pipe, and I just met with a group of them yesterday. I know there is some competition because PVC typically doesn't rust, but there is other problems with it also, so what would you guess would be the usage of PVC compared to metal pipes?

Mr. Proctor. I am not sure what the percentages are exactly, but I can say this, that iron is much more durable than PVC, and their modern techniques virtually eliminate the corrosion for pipe that is installed today.

But even without that, if you look at the track record of iron, as someone mentioned earlier, there was a problem that occurred just the other day for a pipe manufactured in 1860, and that was old cast iron.

Today we have ductal iron that is even stronger and lasts even longer.

Mr. Green. OK. And I know in my area, though, when we see new subdivisions built I almost always see it being built by PVC. Again, because local prices and things like that I guess goes there.

What are the steps that Congress and the EPA can take to ensure that we have the trained workers who need to modernize and maintain our water system? In our district, like I said earlier, we have disadvantaged communities that do not have the resources to invest.

In fact, some of the areas in our district would be called— are colonia, which decades ago was created along the border.

Somebody would go buy, set out a subdivision, but they wouldn't provide any water and sewer, so people would buy a lot, and the only way they could get water is do their own well or a septic tank. But I am also interested in the training for the employees that need to be putting these systems in.

Anybody on the panel? Yes, sir.

Mr. Fletcher. Texas Rural Water Association has circuit riders and technical assistance and training for people like that, for operators that want to learn how to operate a system and get certified. And it is free of charge to these small communities.

Mr. Green. Great. Thank you. I have run out of time. Thank you, Mr. Chairman.

Mr. Shimkus. The gentleman's time has expired. The chair now recognizes the gentleman from Mississippi, Mr. Harper, for 5 minutes.

Mr. Harper. Thank you, Mr. Chairman, and thank you for holding this hearing. I know this is an issue we have looked at for years and continue to be concerned about.

I want to thank each witness for being here and taking time to help us. This is something that as we look at the aging infrastructure in so many of these systems and how we are doing that, and I agree to Mr. Fletcher, the circuit riders in my State of Mississippi have done a remarkable job of helping areas that maybe don't have the resources, and I think that has been a great value across the country where those have been used.

Mr. Vause, if I could ask you a couple of questions. And I know that Mr. Tonko touched on some of this earlier, but I want to try to look a little deeper. I know in your written testimony you emphasize the need for asset management to be encouraged but not mandated. Is there agreement among the industry as to what constitutes good asset management practices?

Mr. Vause. There are basically two models, and those models revolve around five basic concepts. The concepts are more or less solidified between those two models, and so what constitutes good practice really gets to the level of how well you practice each one of those five steps within asset management.

So, I would say generally yes is the answer to that question.

Mr. Harper. OK. But also in that these are sometimes goals or objectives, but how they are met I guess depends upon the resources and determination of each group. Would that be correct?

Mr. Vause. There is. There are policy considerations, considerations that go to what are the necessary levels of service that need to be provided for a particular community. Those are objectives that are set through public policy. There are what are also besides the required levels of service are what are the tolerances that a community has for the degree of risk that they are willing to accept or not accept.

Again, those are public policy choices that are made best at the local level, and so there is no one specific answer.

Mr. Harper. Sure. And of course you are here wearing more than one hat, but on behalf of the American Water Works Association what is that organization doing to encourage or support that better asset management?

Mr. Vause. Yes. We provide through a variety and suite of educational offerings, both in printed materials, in conferences, in workshops, webinars, and so forth, a variety of opportunities for practitioners to be able to learn about these concepts, to see how they are applied both in the United States and elsewhere.

And to bring that information down to the level that allows people from the top executive level down to the plant floor and operators to have the opportunities, the educational opportunities that are necessary to learn how to best apply those practices for their utilities.

Mr. Harper. All right. Well, let's look at where we are right now. If we were talking about what industry or government could do, that might encourage better asset management, does something stand out that you would give us as a takeaway that you want to make sure we don't miss?

Mr. Vause. I think the ability to have the Environmental Protection Agency to be able to monitor these developments and provide materials on a periodic basis to update as time progresses, I think that is an important thing to include in this particular legislation is to ask the administrator to be able to update those on a regular basis and to make them available to all water systems across the United States. I think that is one aspect.

The second aspect that I think is as important is to provide the encouragement through providing a positive incentive to those systems that are interested in securing an SRF loan to be able to reward them for having made positive steps in advancing and adopting those practices at their local utility, not to penalize anyone for not having done so.

But to reinforce through positive rewards, if you will, the ability to work with the agencies and to secure loans so that there is a recognition that advancing these practices leads to good things for utilities.

Mr. Harper. And do you believe you have sufficiently objective criteria to measure that progress?

Mr. Vause. I think there are ways to measure that, and we would certainly be interested in working with the panel here to help identify those specific things that would be able to show measurable progress.

Mr. Harper. Thank you very much. And with that I yield back.

Mr. Shimkus. The gentleman yields back his time. Mr. Chair?

Mr. Tonko. If I might, I know we are rushing off to the briefing for all the House Members.

I just wanted to offer this observation, that everyone is indicating that we need more Federal dollars to address what is a basic core bit of infrastructure that speaks to our needs, individual needs, household needs, and business needs. But if we can find it in our means to provide for 70 billion from the general fund for roads and bridges the FAST Act, I think we need to step up and say, hey, look, this is a hidden infrastructure that cannot be out of sight and out of mind.

We need to do better. We need to prioritize here and not set aside the needs here that should be funded with additional resources from the Federal budget based on recent happenings here in DC.

Mr. Shimkus. And I applaud my colleague for being passionate and committed. So thank you for that.

Seeing there are no further members wishing to ask questions for the panel, I would like to thank you all for coming and also coming early. Again, in my 20 years this is probably the earliest hearing I have been involved with.

Before we conclude I would like to ask for unanimous consent to submit the following document for the record, a letter from the United States Steel Workers. Without objection so ordered.

[The information appears at the conclusion of the hearing.]

Mr. McKinley. And pursuant to committee rules I remind members they have 10 business days to submit additional questions for the record, and I ask that witnesses submit their responses within 10 business days upon receipt of the questions.

And you may get a little bit more since we are so busy this morning, so I think minority counsel warned you all about that previously. Upon receipt of the questions.

Without objection, the subcommittee is adjourned.

[Whereupon, at 10:11 a.m., the subcommittee was adjourned.]

[Material submitted for inclusion in the record follows:]

Statement for the Record U.S. Environmental Protection Agency Committee on Energy and Commerce Subcommittee on Environment

Hearing On "H.R._, Drinking Water System Improvement Act and Related Issues of Funding, Management, And Compliance Assistance under the Safe Drinking Water Act"

May 19, 2017

Chairman Shimkus, Ranking Member Tonko, and Members of the Subcommittee, thank you for the opportunity to provide testimony for today's hearing record regarding the EPA's efforts to support our nation's drinking water infrastructure investments to protect human health.

This statement includes several parts. First, it summarizes the EPA's Drinking Water State Revolving Fund program, which is a significant source of infrastructure funding for our nation's public water systems. Second, it summarizes the EPA's other efforts to understand needs, challenges, and the collaborative work with states, public water systems, and other stakeholders to ensure that our water systems provide clean and safe drinking water to all Americans. Finally, it discusses the concepts of the "Drinking Water System Improvement Act of 2017," which the Subcommittee is discussing today.

Challenges Facing our Nation's Water Infrastructure

Our nation's drinking water infrastructure delivers critical public health protection and serves as a cornerstone for economic development across the country. Some of this infrastructure dates over a century old, which is at or beyond its useful life. The EPA's 2011 Drinking Water Needs Survey identified nearly $384 billion in capital improvement needs, eligible for the Drinking Water State Revolving Fund, to keep pace with the aging of this critical drinking water requirement over the next 20 years. These investments comprise pipe and other components of drinking water distribution systems, as well as thousands of treatment plants, storage tanks, and other key assets to ensure the public health, security, and economic well-being of our cities, towns, and communities. Implementing the projects that are needed to maintain and upgrade our existing drinking water infrastructure will remain an essential strategy for protecting the public health in America's communities in the years ahead.

Drinking Water State Revolving Fund

Through the Drinking Water State Revolving Fund (DWSRF), community drinking water systems of all sizes are supported with assistance to maintain the essential components and functions of these systems. Established by the 1996 Safe Drinking Water Act (SDWA) Amendments, the DWSRF is one important tool available to states and local water systems as they seek to address the challenge of continuing to provide safe drinking water. The program creates efficient and sustainable financing programs uniquely tailored to each state's special circumstances, making it a highly successful state-federal partnership and an important complement to the new Water Infrastructure Finance and Innovation Act (WIFIA) loan program.

The EPA provides capitalization grants to the state DWSRF programs as an investment in the nation's infrastructure. By contributing an additional 20 percent of what the EPA provides, states further enhance the size and effectiveness of the program. Twenty-two states leverage their program on the tax-exempt debt market to increase their lending capacity. They make loans at below-market rates, at an average of two percentage points below market over the last several years. Often the result is a substantial interest savings for communities, providing the equivalent to a grant covering approximately 20% of the cost of a project. States even have the flexibility to charge no interest over the life of a loan.

The programs operate on the basis of cost reimbursement. Even though a grant is made directly to a state by the EPA, no funds leave the Treasury until costs are incurred. Through 2016, a total of nearly $32.5 billion in assistance has been provided by the 51 DWSRF programs to more than 13,000 projects across the country. Over the last three years, the 51

DWSRF programs have provided on average $2.25 billion per year to communities to finance about 770 projects each year, including assistance to non-infrastructure, capacity-building, and prevention-focused set-asides.

The fiscal year 2018 President's Budget provides robust funding for critical drinking water infrastructure investment. It furthers a commitment to infrastructure repair and replacement, which would allow states, municipalities, and private entities to continue to finance high priority infrastructure investments that protect human health. The President's Budget also includes a total of $2.3 billion for the State Revolving Funds (including both the DWSRF and the Clean Water SRF), a $4 million increase over the 2017 annualized Continuing Resolution level.

PROTECTING PUBLIC HEALTH AND THE ENVIRONMENT WITH DWSRF

Priority for DWSRF assistance is given to systems facing an immediate threat to public health, systems with infrastructure investment needs to comply with SDWA health standards, and systems most in need on a per-household basis according to state affordability criteria. Reflecting these priorities, about 40% of projects involve treatment upgrades, 40% involve rehabilitation or replacement of distribution pipes, 10% involve source water, and 10% involve improvements to finished water storage. Repayments, a significant feature of the SRFs, are recycled back into the program to provide a source of ongoing funding for additional drinking water projects. Through mid-2016, nearly $8 billion in principal and interest has been returned to the DWSRFs by borrowers.

Additionally, states have the ability to leverage federal grant awards through the sale of tax-exempt bonds. A very basic example of bond leveraging is a state that receives a $10 million annual capitalization grant. Using its stream of repayments as security, the state might issue $20 million in bonds, "leveraging" its $10 million capitalization grant to get $30 million in lending capacity. The net proceeds of these bonds have provided over $7 billion in additional funding for critical projects.

States have the authority, under the DWSRF, to use a portion of their capitalization grants for additional subsidization in the form of principal forgiveness or grants. This valuable authority allows provision of critical resources to the neediest communities unable to afford SRF loans. To date, states have provided nearly $3 billion in additional subsidy to state-identified disadvantaged communities.

Small water systems (those serving 10,000 or fewer persons) have received 71% of the total number of assistance agreements made over the program's history, but account for 35% of all dollars awarded for assistance. In contrast, the largest cities (those serving more than 100,000 persons) have received 7% of the number of assistance agreements. Because of the size and complexity of these large systems, their agreements account for 27% of all dollars awarded for assistance. The DWSRF has successfully established a record of addressing varying water system needs across our nation's communities, both small and large.

A significant feature of the DWSRF is the flexibility it provides states to use up to 31% of each capitalization grant for a variety of set-asides. The set-asides help states fund administration of the DWSRF, provide technical assistance to small systems, advance the core public health protection mission of state drinking water programs, and support system-level efforts to enhance efficiency and performance.

CAPACITY DEVELOPMENT AND OPERATOR CERTIFICATION PROGRAMS

While DWSRF funds play an important role in addressing the nation's infrastructure needs, the 1996 amendments to SDWA created the Capacity Development and Operator Certification programs. In implementing these programs, the EPA is also playing a broader role in working to ensure that investments by federal, state, and local governments, as well as the private sector, will yield the public health protections they are intended to support. Toward this end, through the DWSRF program and other efforts, the EPA works with states and local communities to emphasize the importance of asset management and capacity development at the state and local level. The EPA is working with partners across the water sector and beyond to provide the knowledge and tools to ensure that the investments we make in our water infrastructure yield resilient and sustainable public health protections. This goal can be achieved through partnerships with states, tribes, local governments, and water systems to develop and maintain technical, managerial, and financial capacity, as well as by promoting professional development in the water sector in order to ensure that there is a pool of qualified water professionals to meet current and future needs. The EPA is targeting its resources toward helping systems achieve results in the following areas:

- Promoting an asset management framework that ensures the right investments are made at the right time;
- Promoting water system partnerships that create opportunities to improve service and public health protection, reduce costs, and address future needs;
- Promoting infrastructure financing and providing options to pay for water infrastructure needs, including through full-cost pricing; and
- Promoting investment in a strong water workforce through capacity development, operator certification, and knowledge sharing, recruitment, and training in the water sector.

The Drinking Water System Improvement Act of 2017

The Administration has not taken a position on the Drinking Water System Improvement Act of 2017, but the EPA appreciates the opportunity to provide information relevant to the important issues that this bill would address. These issues include the DWSRF; the State Public Water System Supervision (PWSS) program grants; efforts funded by the DWSRF set-asides, such as asset management and water system partnerships; and demonstration of compliance with federal cross-cutting requirements, each of which plays a critical role in protecting public health in communities across the nation.

As designed within SDWA, the states' PWSS programs are the foundation of the implementation of SDWA by the states, and the federal PWSS grant assists in the successful operation of state programs. The PWSS program and the DWSRF set-asides are fundamental to ensuring effective implementation of state drinking water programs. The PWSS program supports conducting sanitary surveys, providing technical assistance to public water systems, developing and maintaining state drinking water regulations, ensuring that public water systems provide information to their consumers, and performing other core program implementation functions. The DWSRF set-asides support activities necessary to ensure safe and affordable drinking water such as asset management, water system partnerships, training and technical assistance, financial management and rate studies, and source water and wellhead protection. Asset management; water system partnerships, including consolidation; and source water protection are important tools to develop and maintain sustainable drinking water systems. As noted earlier in this statement, the agency works with states, tribes, local governments, water systems, and other water sector stakeholders to support the development of these programs. With this wide range of potential activities, the DWSRF set-asides have become an important source of funding for state drinking water programs.

Additionally, providing flexibilities to states in the use of DWSRF funding can also support states and systems in ensuring the protection of public health. The use of additional subsidization and extended loan periods are important options to help small and disadvantaged communities improve and maintain sound drinking water infrastructure. In the application of these flexibilities, it is important to ensure that the states are able to manage the funds in perpetuity.

Conclusion

The EPA's DWSRF program is focused on actions and funding to achieve compliance with environmental and public health standards. Addressing these challenges will require effort from the EPA, states, communities, and other partners. It will require us to use more innovative and sustainable tools to solve these significant challenges. We look forward to working with Members of the Subcommittee, our federal and state colleagues, and our many partners, stakeholders, and citizens who are committed to continuing our progress in providing clean and safe drinking water to all Americans.

Congress of the United States
House of Representatives
COMMITTEE ON ENERGY AND COMMERCE
2125 Rayburn House Office Building
Washington, DC 20515–6115
Majority (202) 225–2927
Minority (202) 225–3641

June 9, 2017

The Honorable Scott Pruitt
Administrator
U.S. Environmental Protection Agency
1200 Pennsylvania Avenue, N.W.
Washington, DC 20460

Dear Administrator Pruitt:

Thank you for submitting a statement for the record for the Subcommittee on Environment's Friday, May 19, 2017 hearing entitled "H.R._, Drinking Water System Improvement Act and Related Issues of Funding, Management, and Compliance Assistance under the Safe Drinking Water Act."

Pursuant to the Rules of the Committee on Energy and Commerce, the hearing record remains open for ten business days to permit Members to submit additional questions for the record, which are attached. The format of your responses to these questions should be as follows: (1) the name of the Member whose question you are addressing, (2) the complete text of the question you are addressing in bold, and (3) your answer to that question in plain text

To facilitate the printing of the hearing record, please respond to these questions with a transmittal letter by the close of business on Friday, June 23, 2017.

Thank you again for your assistance to the Subcommittee.

Sincerely,

John Shimkus
Chairman
Subcommittee on Environment

cc: The Honorable Paul Tonko, Ranking Member, Subcommittee on Environment

EPA Responses to Questions for the Record

Honse Committee on Energy and Commerce, Subcommittee on Environment

May 19, 2017, Hearing on ""H.R._, Drinking Water System Improvement Act and Related Issues of Funding, Management, and Compliance Assistance under the Safe Drinking Water Act."

THE HONORABLE JOHN SHIMKUS

What do You Consider the Core Mission and Programs of the Agency?

The mission of the EPA is to protect human health and the environment. In carrying out its mission, the EPA works to ensure that all Americans are protected from exposure to hazardous environmental risks where they live, learn, work, and enjoy their lives.

Under Administrator Pruitt, the EPA is building on the agency's progress to date by focusing on three core philosophies for carrying out the EPA's mission:

- Rule of law: Administering the laws enacted by Congress and issuing environmental rules tethered to those statutes, relying on agency expertise and experience to carry out congressional direction and to ensure that policies and rules reflect common sense and withstand legal scrutiny.
- Cooperative federalism: Recognizing the states and tribes, as applicable, as the primary implementers and enforcers of many environmental laws and programs, and partnering with them to engender trust and maximize environmental results to protect human health and environment.
- Public participation: Fulfilling obligations to conduct open and transparent rulemaking processes, engaging with and learning from the diverse views of the American public, and addressing stakeholder input on the impacts of rules on families, jobs, and communities.

How does the DWSRF and SDWA Fit into a Back to Basics Strategy?

Administrator Pruitt's "Back-to-Basics Agenda" reflects his efforts to refocus the EPA on its intended mission, return power to the states, and create an environment where jobs can grow. The agenda focuses on the three E's:

Environment: Protecting the environment
Economy: Sensible regulations that allow economic growth
Engagement: Engaging with state and local partners.

A priority for the agency is modernizing the outdated water infrastructure on which the American public depends. While most small systems consistently provide safe and reliable drinking water, many small systems face challenges with aging infrastructure, increasing operational costs, and decreasing rate bases. The President's FY 2018 budget provides funding for critical drinking and wastewater projects. These funding levels support the President's commitment to infrastructure repair and replacement and would allow states, municipalities, and private entities to finance high-priority infrastructure investments. The FY 2018 budget includes $2.3 billion for the State Revolving Funds and $20 million for the Water Infrastructure Finance and Innovation Act (WIFIA) program. Under WIFIA, the EPA could potentially provide up to $1 billion in credit assistance, which, when combined with other funding sources, could spur an estimated $2 billion in total infrastructure investment. This makes the WIFIA program credit assistance a powerful new tool to help address a variety of existing and new water infrastructure needs.

The EPA will continue to partner with states, drinking water utilities, and other stakeholders to identify and address current and potential sources of drinking water contamination. These efforts are integral to the sustainable infrastructure efforts as source water protection can reduce the need for additional drinking water treatment and associated costs. As progress has been made, work remains for existing and emerging issues.

THE HONORABLE FRANK PALLONE, JR.

Buy America

1. Is it the policy of this Administration to support Buy America requirements on projects financed by Drinking Water State Revolving Fund loans? Yes.

Asset Management

Section 3 of the Drinking Water System Improvement Act discussion draft amends section 1420(c) of the Safe Drinking Water Act, which conditioned receipt of SRF capitalization grants on the development and implementation of state capacity development strategies. Current law does not include a requirement to periodically revise those capacity development strategies, and the discussion draft does not create such a requirement. Despite this, the discussion draft adds a new requirement for the content of those plans.

2. Have all states developed capacity development strategies under this section? Yes, all states with Safe Drinking Water Act primacy and Puerto Rico have developed capacity development strategies.
3. Have states periodically revised these strategies? Yes, states revise their strategies to incorporate new initiatives and/or programmatic changes. During the revision process, states seek public and stakeholder input on strategy revisions.
4. Has EPA required states to periodically revise these strategies? No. The EPA does not require strategy revisions but suggests states review and update their strategies, and many states have done so. The EPA supports state capacity development activities through information sharing on best practices, and through development of tools and resources.
5. Would the language in section 3 create a requirement for states to revise and resubmit these strategies? The language may require some states to revise and resubmit their capacity development strategy. Some states may already have asset management included in their strategy.
6. The 1996 SDWA amendments provided 4 years for states to develop the capacity development strategies - how much time would be provided under section 3 for states to revise these strategies before they are penalized with decreased funding? The EPA would consult with Congress, states, and key stakeholders, to identify an appropriate time to allow for strategy revisions.

Source Water Protection

Section 6 of the Drinking Water Systems Improvement Act discussion draft amends the source water protection provisions in the Safe Drinking Water Act in two ways. First, it removes the fiscal year limitation on the use of SRF capitalization grants by states for source water protection. Second, it bars the use of those funds for costs arising from requirements under the Federal Water Pollution Control Act.

7. Does this section make additional funding available for source water protection activities? States have not been using the full 15% of the DWSRF set-aside under SDWA Section 1452(k). Therefore, most states would have the ability to take additional funds in this section for source water assessments if they choose to. States can make source water assessments a regular and ongoing activity, since they may take this set-aside every year.
8. If these activities are to be funded from current capitalization grants, do many states have surplus funds to direct towards source water protection activities? We would interpret the changes made by this legislation to apply to future capitalization grants,

and thus currently unspent funds from prior capitalization grants would not be used for this purpose.

9. Current funding allotments are based on EPA' s needs assessment - does that assessment incorporate source water protection costs? No. The EPA's Drinking Water Infrastructure Needs Survey includes only the rehabilitation or replacement of existing infrastructure or installation of eligible new infrastructure, such as water treatment plant components or a drinking water intake. It does not include source water protection activities.
10. Is the limitation on using funds for costs arising from requirements under the Federal Water Pollution Control Act a new limitation? If no, is the additional language needed? If yes, what will the impact of this limitation be? Existing SDWA language requires that funds for source water protection can only be used to fund voluntary, incentive-based mechanisms. The EPA is not aware of any funding under 1452(k) that has gone to support compliance with Federal Water Pollution Control Act (Clean Water Act) requirements.
11. Who would be responsible under this language for determining what source water protection activities can be funded under section 1452(k)(l) versus what costs arise from requirements under the Federal Water Pollution Control Act? The state DWSRF program is responsible for complying with statutory requirements. The EPA reviews states' Intended Use Plans, Annual Reports, and financial documentation to ensure that DWSRF funds are spent appropriately. If this provision were enacted, the EPA would issue guidance, if necessary, to explain or clarify the statutory requirements as part of the agency's role in overseeing state DWSRF programs.
12. If a state used funds under this section for source water protection activities that contributed to compliance under the Federal Water Pollution Control Act, would the state be penalized? What would the penalties be, and how would they be enforced? See response to Question 11 above. Given the potential for ancillary benefits from source water protection activities for Clean Water Act compliance, the EPA would need to carefully consider how best to implement this provision. The EPA website provides resources to assist with source water protection and states, "Preventing source water contamination is preferable to remedying its negative effects." The website also says that "Preventing source water contamination can be less costly than remedying its effects."
13. Do these statements still reflect the position of the EPA with regards to source water protection? Yes.

Cross-Cutting Requirements

Section 8 of the Drinking Water System Improvement Act discussion draft grants the EPA Administrator to accept demonstrations of compliance with state or local laws as a demonstration of compliance with "federal cross-cutting requirements" that are equivalent. That section defines the term "Federal cross-cutting requirement" as a federal requirement that would be redundant with a requirement of an applicable state or local law.

14. This section introduces two different standards for comparing federal and state requirements - first that they are "equivalent" and second that they are "redundant." Would the EPA interpret these standards as meaning the same thing? The EPA would interpret these terms in a complementary way. If the federal requirement is the same as an existing state requirement, then it would be considered to be duplicative or redundant. In some cases, a state may have a requirement that EPA believes achieves the same result as the federal requirement. In such a case, EPA could determine that the state requirement is equivalent to that of the federal action.
15. What cross-cutting [requirements] do you anticipate would be covered by this section? Does this apply to demonstrations to be made to the EPA by states receiving capitalization grants under the SRF? Does this apply to demonstrations to be made to states by water systems receiving loans under the SRF? Additional information on the cross-cutting federal authorities potentially applicable to the DWSRF program are outlined in the EPA's cross-cutter handbook at https://www.epa.gov/sites/production/files/2015-08/documents/crosscutterhandbook.pdf. A number of these cross-cutting provisions could be covered by this section. The demonstration of compliance with a particular authority would be made to the EPA by states receiving capitalization grants or water systems receiving loans under the SRF.
16. Under the language, the Administrator determines whether a demonstration is "equivalent" but the definition seems to be ambiguous as to who determines what requirements are "redundant." How would you interpret this ambiguity? The EPA would implement the provisions similarly. If the Administrator determines that a state provision is equivalent to a federal requirement or if the Administrator determines that a federal requirement is redundant with a state requirement, then the EPA would allow a demonstration of compliance with the state requirement to count as compliance with the federal requirement.

Lead and Copper Rule Long-Term Revisions

17. Last year, EPA testified before the Committee that the long-term revision of the Lead and Copper Rule (LCR) was expected to be finalized in 2017. Is EPA still on track to publish a revised LCR in the coming months? If not, what has changed? Protecting children from exposure to lead is a top priority for the EPA. The agency has conducted extensive engagement with stakeholder groups and the public to inform potential revisions to the LCR. The EPA is carefully evaluating the recommendations from these

groups and is giving extensive consideration to the national experience in implementing the rule as well as the experience in Flint, Michigan, as we develop proposed revisions to the rule. The EPA must also consider the potential impact of these regulatory revisions on the thousands of communities across the country that will have to implement these requirements. The EPA plans to provide additional information regarding its rulemaking timeline in the unified agenda later this summer.

18. How would cuts to EPA funding in the President's budget impact your ability to finalize revisions to the Lead and Copper Rule? Protecting children from exposure to lead is a top priority for the EPA. We will continue to assure that resources are available to improve public health protections under the Lead and Copper Rule.
19. How would cuts to the Office of Enforcement and Compliance Assurance (OECA) in the President's budget impact your ability to enforce current requirements under the lead and copper rule? The EPA will continue to coordinate with states, tribes, and territories to enforce not only the Lead and Copper Rule, but all SDWA national primary drinking water regulations. The EPA will continue its work with our co-regulators to ensure that owners/operators of public water systems address noncompliance in a timely manner, prioritizing those systems with the most serious or repeated violations.

Flint Response

20. How would cuts to EPA funding in the President's budget impact your ability to provide guidance and technical assistance to the community of Flint, Michigan? In March, the EPA awarded a $100 million grant to the Michigan Department of Environmental Quality to fund drinking water infrastructure upgrades in Flint. The EPA will continue to work with the State of Michigan, the City of Flint, and other federal agency partners to improve the City's public water system. More generally, the agency will continue to work with states, including the state of Michigan, to implement requirements for all national primary drinking water regulations and to ensure that drinking water systems, including the City of Flint, install, operate, and maintain appropriate levels of treatment and effectively manage their distribution systems. For instance, the EPA will continue to focus on working with states to optimize corrosion control treatment to minimize exposure to lead. The EPA will also continue to focus on small systems by strengthening and targeting financial assistance, in coordination with state infrastructure programs, to support rehabilitation of the nation's infrastructure. The agency also will look for ways to promote partnerships among water systems to build capacity and work with states and tribes, as well as with utility associations, third-party technical assistance providers and other federal partners, to

promote the sustainability practices that are the foundation for building technical, managerial, and financial capacity.

21. How would the cuts impact your ability to continue to monitor chlorine levels biweekly and collect sequential samples for lead assessment on a bimonthly basis? The EPA concluded regular chlorine monitoring in Flint at the end of 2016 because the city began assessing chlorine levels more frequently at additional locations throughout the city in early 2017. The EPA's last round of sequential sampling for lead took place in November 2016. As such, the EPA is no longer conducting biweekly chlorine monitoring or sequential lead sampling in Flint, so any changes in the EPA's budget would not affect these past sampling activities.
22. In December, EPA agreed with recommendations from the EPA Office of the Inspector General (OIG) to issue updated guidance through OECA on emergency authority under Section 1431 of the Safe Drinking Water Act. That guidance is due to be issued in November 30, 2017. Do you still anticipate issuing that guidance by November 30, 2017? How will proposed budget cuts for OECA affect the issuance of that guidance? Yes, in accordance with the OIG's October 2016 Management Alert (Alert) regarding EPA authority to issue emergency orders to protect public health, OECA still plans to issue updated SDWA Section 1431 guidance and train all relevant EPA drinking water and water enforcement staff and management on Section 1431 by the November 30, 2017, deadline. Given the scope of drinking water issues and resources available, the EPA continually works to prioritize matters and protect public health. In this regard, OECA recognizes the importance of the issues raised in OIG's Alert and, thus, has maintained efforts to update our SDWA Section 1431 guidance and conduct training.
23. EPA also agreed in December with the OIG's recommendation to train all relevant EPA drinking water and water enforcement staff on Section 1431 authority by November 30th, 2017. Do you still expect to complete that training by November 30, 2017? How will proposed budget cuts for OECA affect that training? OECA still plans to train all relevant EPA drinking water and water enforcement staff and management on Section 1431 by the November 30, 2017, deadline. Please see our response to Question 22 above.

Board of Scientific Counselors

24. Last month, EPA dismissed many members of the Board of Scientific Counselors (BOSC). Please provide the full list of members who were terminated, as well as those who remain. Similar to other federal advisory committees, BOSC members are appointed to serve a three-year term as a Special Government Employee (SGE), which can be renewed once. On April 28, the three-year terms expired for nine members of the BOSC Executive Committee (names provided below) and their terms were not

renewed. On May 25, the EPA published a Federal Register Notice soliciting public nominations for members of the BOSC. On June 19th, BOSC members whose terms will be expiring in August were also informed that their terms would not be renewed. Those members whose terms had expired or will be expiring shortly were informed that they could reapply for consideration during this nomination period.

Members Whose First Terms Expired

Viney Aneja	Courtney Flint	Tammy Taylor
Sandra Smith	Shahid Chaudhry	Ponisseril Somasundaran
Robert Richardson	Paula Olsiewski	Gina Solomon

Members Whose Second Terms Ended (Members cannot Serve More than Two Terms)

John Tharakan	Susan Cozzens
Earthea A. Nance	Diane Pataki

Members Who Resigned

Peter Meyer   Carlos Martin   Elizabeth Corley

Members Who Remain (Those Whose Terms Expire in August 2017 are Marked with a *)

Deborah Swackhamer	Andrew Dannenberg*	Monica Schoch-Spana*
James Galloway	Richard Feiock*	Michael Wichman
Joseph Rodericks*	Elena Irwin*	Lance Brooks
Leslie Rubin	Matthew Naud*	Katrina Waters*
Jeffrey Arnold*	Mike Steinhoff*	Chris Gennings
Elena Craft*	Deborah Reinhart*	Dale Johnson*
Charlette Geffen*	James Kelly	Rebecca Klaper*
Donna Kenski*	Scott Ahlstrom*	Kyle Kolaja
Patrick Kinney*	Bruce Aylward*	Jerzy Leszczynski*
Myron Mitchell*	Lawrence Baker*	Jennifer McPartland*
Constance Senior*	Inez Hua*	James Stevens*
Art Werner*	John Lowenthal*	Donna Vorhees
Jinhua Zhoa*	Shane Snyder*	Clifford Weisel*
Louie Rivers*	Andrew DeGraca*	Mark Wiesner*

Todd BenDor*	Edward Hackney*	Paloma Beamer*
Robert Cervero	Edwin Roehl*	

25. Which members of the BOSC Safe and Sustainable Water Resources Subcommittee have been terminated? Of the nine members whose first three-year term expired and was not renewed, only one was a member of the Safe and Sustainable Water Resources Subcommittee. Dr. Shahid Chaudhry served as the Vice Chair of the SSWR subcommittee since 2014.
26. EPA is currently seeking nominations to replace the terminated BOSC members. According to EPA's website, you are currently seeking nominations for scientists with expertise in drinking water treatment, nutrient management, climate change, risk assessment, and other drinking water safety concerns. How will these expertise gaps affect your ability to seek advice from the BOSC until the positions are filled? The nine BOSC members who were not renewed for a second term represent a wide spectrum of expertise and were members of multiple BOSC subcommittees, including the Safe and Sustainable Water Resources subcommittee. Due to the time required for new solicitation and vetting of applicants, the previously scheduled BOSC meetings through early fall have been postponed. Once the BOSC is reconstituted, with the appropriate expertise, the review of the EPA's research will begin as soon as possible. However, the EPA's research will continue in the interim.
27. Will the BOSC vacancies affect the timeline for revisions of the Lead and Copper Rule or any other rulemaking under the Safe Drinking Water Act? Vacancies in the BOSC will not impact development of proposed revisions to the Lead and Copper Rule, nor will it impact other SDWA rulemaking activities because the BOSC has not been charged to review scientific products associated with current SDWA rulemakings.
28. What is your timeline for filling the BOSC vacancies? The EPA anticipates that these vacancies will be filled by late 2017 or early 2018.
29. What opportunities for public participation will be provided in the selection of new BOSC members? The EPA's outreach plan for developing a diverse pool of BOSC nominees includes an open solicitation of potential candidates, which is published in the Federal Register, as well as updates on the EPA's website (https://www.epa.gov/bosc/invitation-nominations-bosc-executive-committee-and-subcommittees). Further outreach is conducted through multiple professional associations and organizations to encourage nomination of a broad range of candidates.

Perfluorinated Compounds (PFCs)

30. Has EPA worked with the Department of Defense (DOD), Air Force, or Navy to respond to the emerging contamination of drinking water caused by PFCs on and around DOD installations?
31. Please explain any information sharing, technical assistance, or coordinated response that has occurred between EPA and DOD. Yes. At both the Regional and Headquarters levels, the EPA is regularly engaged in discussions with DOD and its component services on perfluorinated compounds. For example, EPA Headquarters has ongoing quarterly meetings with each of the DOD components to discuss salient topics such as PFAS. Further, EPA Headquarters and Regions meet biannually with DOD components by inviting them to participate in the EPA's Federal Facility Leadership Council. FFLC meetings allow Regional managers to discuss site specific issues, such as PFAS investigations and responses, within a broader context with the National Program managers of both EPA and, through EPA's invitation, DOD. The EPA also briefed DOD along with other federal agency partners in spring 2016 regarding the EPA's final health advisories for PFOA and PFOS. At National Priorities List (NPL) DOD sites, the EPA is actively engaged to help ensure a timely, protective response to perfluorinated compounds contamination of drinking water in both the Regions and Headquarters. Cleanup agreements between the EPA and Federal agencies at such NPL sites require the Federal agency to investigate and remediate hazardous substances, pollutants, and contaminants, and PFCs are pollutants and contaminants. When appropriate, the EPA has used enforcement authority to address PFOA/PFOS contamination. For instance, in 2014 and 2015, the EPA issued three Safe Drinking Water Act orders to two DOD components when the EPA determined that there may be an imminent and substantial endangerment. These orders require actions such as the provision of bottled water (where drinking water exceeded the lifetime health advisory), off-site residential well sampling, and the treatment of contamination at wells in order to protect public supply wells and restore the underlying aquifer. Such work is ongoing, and the DOD components are currently in compliance with the orders.
32. Has EPA encouraged states to notify firefighting departments, civilian airports, or other organizations that may have utilized or stored aqueous film forming foam about the risks of PFC contamination? The EPA has not specifically encouraged states to notify firefighting departments, civilian airports, or other organizations that may have utilized or stored aqueous film-forming foam about the risks of PFC contamination, but the EPA has more generally advised the public of these risks through its ongoing work with state partners, on the EPA's website, and in other ways. Information on the EPA's actions regarding PFC contamination is available at https://www.epa.gov/assessing-and-managing-chemicals-under-tsca/and-polyfluoroalkyl-substances-pfass-under-tsca.

33. Has EPA considered supplying states or public water systems with a list of best available technologies to treat PFC drinking water contamination? Yes. The EPA's 2016 health advisory for PFAS provided information on options available to drinking water systems to lower concentrations of PFOA and PFOS in their drinking water supply. Public water systems can treat source water with activated carbon or high pressure membrane systems (e.g., reverse osmosis) to remove PFOA and PFOS from drinking water. Treatment technology information is available in the Drinking Water Health Advisory documents at: https://www.epa.gov/ground-water-and- drinking-water/supporting-documents-drinking-water-health-advisories-pfoa-and-pfos.
34. Is EPA currently considering issuing a national drinking water standard on any perfluorinated compound? The EPA included PFOA and PFOS on the fourth Contaminant Candidate List (CCL 4) and is evaluating these contaminants to determine if they meet the three SDWA regulatory determination criteria in Section 1412(b)(1)(A):
    (i) may have an adverse effect on the health of persons;
    (ii) is known to occur or there is a substantial likelihood that it will occur in public water systems with a frequency and at levels of public health concern;
    (iii) In the sole judgment of the Administrator, regulating the contaminant presents a meaningful opportunity for health risk reductions for persons served by public water systems.

    The EPA plans to make determinations to regulate or not regulate at least 5 contaminants from the fourth Contaminant Candidate List by January 2021. The EPA expects to publish preliminary regulatory determinations for public comment in 2019.

WIIN Act Authorizations

In 2016, Congress authorized three new grant programs to promote safe drinking water in the Water Infrastructure Improvements for the Nation Act (Public Law No: 114-322)

- Lead service line replacement grant program authorized at $60 million annually from FYl7 to FY21;
- Assistance for small and disadvantaged communities grant program authorized at $60 million annually from FY17 to FY21; and,
- Voluntary school and child care lead testing grant program authorized at $20 million annually from FY17 to FY21.

35. Please provide an update on EPA's implementation of these three programs. The omnibus spending bill enacted in May to fund the government through the end of September included water-related categorical grants, which support state and tribal

programs, and maintain FY16 enacted levels. While Title II of WIIN authorizes several new grant programs (such as Sections 2104, 2105, and 2107), Congress has not appropriated funding for these programs. The EPA is preparing for implementation should appropriations be made available. In the meantime, the EPA continues to partner with states, drinking water utilities, and other stakeholders to implement and support drinking water programs.

36. The President's FY18 Budget Request did not include funding for these programs. Is EPA prepared to award grants in FY18, either through a reprogramming of existing funds or an appropriation from Congress? The EPA is preparing for implementation should appropriations be made available. In the meantime, the EPA continues to partner with states, drinking water utilities, and other stakeholders to implement and support drinking water programs.

Chlorpyrifos

In April 2016, EPA published a revised chlorpyrifos drinking water assessment and found "potential exposure to chlorpyrifos or chlorpyrifos-oxon in finished drinking [water] based on currently labeled uses." Chlorpyrifos is a dangerous pesticide that causes serious neurodevelopmental harm in infants and children, including delayed mental development, attention problems, autism spectrum disorders, and intelligence decrements. EPA itself found these effects in a rigorous risk assessment vetted by the Science Advisory Panel. Despite these clear findings, EPA recently denied a petition to ban chlorpyrifos.

37. Given EPA's shocking decision to allow continued use of chlorpyrifos, what will be done to address and eliminate the risk of chlorpyrifos exposure from drinking water? Following a review of comments on both the November 2015 proposed tolerance revocation and the November 2016 notice of data availability, which included updated human health and drinking water assessments, the EPA concluded that the science addressing neurodevelopmental effects remains unresolved and that further evaluation of the science during the remaining time for completion of registration review is warranted to achieve greater certainty regarding the risk of adverse neurodevelopmental effects at current levels of human exposures to chlorpyrifos. Accordingly, on March 29, 2017, the EPA denied the citizen petition seeking revocation of chlorpyrifos tolerances, concluding that the appropriate course of action is to take steps to come to a clearer resolution on the potential risks of chlorpyrifos before completing the registration review or any associated tolerance action. Under the Federal Insecticide, Fungicide, and Rodenticide Act (FIFRA), the EPA must complete registration review by October 1, 2022.

Climate Change

On June 1, 2017, President Trump announced his intention to withdraw the United States from the Paris Climate Accord, imperiling our progress in the fight against climate change. This followed a May 25, 2017 briefing provided to the Energy and Commerce Committee on the President's FY 2018 budget, at which an EPA representative stated that climate change is "no longer a priority" for this administration, and that the agency's focus would be on issues impacting human health. But climate change has significant and undeniable impacts on human health, including on the safety of drinking water.

38. EPA's own website states that harmful algal blooms (HABs) might "occur more often, in more water bodies, and be more intense" because of climate change, and acknowledges that the unregulated microcystin toxins from the blooms "endanger human health." How will this change in priorities affect efforts by EPA to address the risks to drinking water safety and human health from the impacts of harmful algal blooms? How will the potential eventual withdrawal from the Paris Climate Accord impact the public health risks from harmful algal blooms? Many sources of drinking water face risks for developing harmful algal blooms and cyanotoxin occurrence. In accordance with the Algal Toxin Risk Assessment and Management Strategic Plan for Drinking Water submitted to Congress in November 2015, the EPA will continue to work with our State and local partners to:
    - Assess the frequency and level of cyanotoxin occurrence in public water systems nationwide under the unregulated contaminant monitoring rule;
    - Improve understanding of the human health effects of current and emerging cyanotoxins;
    - Provide technical assistance to water systems to manage HABs, monitor and treat cyanotoxins in drinking water;
    - Support efforts for source water protection and nutrient reduction strategies at the watershed scale; and
    - Improve scientific understanding of HABs and cyanotoxin production to better predict their occurrence.

The EPA has developed a number of tools that public water systems can use to reduce the risks of cyanotoxins occurring in finished drinking water. These tools are available on the EPA's website at https:// www.epa.gov/ground-water-and-drinking-water/cyanotoxin-tools-public-water-systems.

39. Climate change also threatens the availability and reliability of drinking water sources, through more frequent droughts, floods, and extreme weather events. How will EPA's change in priorities affect efforts to protect and adapt our drinking water infrastructure

to droughts, floods, and extreme weather events? How will the potential eventual withdrawal from the Paris Climate Accord impact the public health risks from droughts, floods, and extreme weather events? The EPA has an important mission through its homeland security responsibilities as the Sector Specific Agency for Water to facilitate the protection of the nation's critical water infrastructure from all hazards, including droughts and other severe weather events. The EPA will continue to enhance the resilience of water systems through an extensive array of programmatic tools, such as WIFIA, which will make financing available to drinking water systems for infrastructure improvements, for example, to address drought prevention, reduction, or mitigation projects.

40. Climate change also threatens drinking water sources through sea level rise and saltwater intrusion into aquifers. How will EPA's change in priorities and the potential eventual withdrawal from the Paris Climate Accord affect the public health risk from the effects of sea level rise and saltwater intrusion on drinking water? The EPA has an important mission through its homeland security responsibilities as the Sector Specific Agency for Water to facilitate the protection of the nation's critical water infrastructure from all hazards, including saltwater intrusion into aquifers. The EPA will continue to enhance the resilience of water systems through an extensive array of programmatic tools, such as WIFIA, which will make financing available to drinking water systems for infrastructure improvements, for example, to address brackish or seawater desalination, aquifer recharge, alternative water supply, and water recycling projects.
41. Climate change also threatens the safety of drinking water because higher temperatures can lead to greater leaching of lead from pipes and plumbing fixtures; proliferation of viruses and bacteria in our drinking water distribution systems; and increases in concentrations of pollutants such as ammonia. How will EPA's change in priorities and the potential eventual withdrawal from the Paris Climate Accord affect the public health risk from rising temperatures? The EPA has promulgated a number of national primary drinking water regulations to address contaminants, such as lead and pathogens. The agency will continue to work with states to implement requirements for all drinking water regulations to ensure that water systems install, operate, and maintain appropriate levels of treatment and effectively manage their distribution systems.
42. A May 17, 2017 memorandum from EPA acting CFO David Bloom notes an adjustment "in the Climate Protection Program reflecting reduced activity" but does not specify a dollar amount. What is the dollar amount associated with this budget reduction and how will this reduction impact implementation of climate programs? The May 17, 2017, memorandum represents an internal EPA planning document for developing an operating plan after enactment of the Consolidated Appropriations Act, 2017. There was a minimal adjustment to that account that is within the agency's

reprogramming limitations. The agency submitted its FY 2017 Enacted Operating Plan to Congress on June 5.

43. The President's proposed FY 2018 budget cuts EPA's budget by nearly $2.6 billion - an overall 31 percent reduction- and includes extreme cuts to key public health and environmental programs, such as grants and programs for state and tribal air quality, diesel emission reductions, and lead safety. What analysis, if any, has the agency conducted to assess the impact of these reductions on human health? What did the analysis conclude? The agency's FY 2018 budget lays out a comprehensive back-to-basics and foundational strategy to maintain core environmental protection with respect to statutory and regulatory obligations. The agency's FY 2018 Budget in Brief and associated Congressional Justification describe the budget and the EPA's programs in greater detail. The Congressional Justification includes provisional FY 2018 performance measures that provide more detail on the specific activities the EPA would undertake in FY 2018 to protect human health and the environment. On the EPA website entitled "Addressing Climate Change in the Water Sector," several links that have previously provided valuable information to affected communities are now described as "being updated". Examples include "Explore Your Climate Region", "Climate Impacts on Water Resources", "Climate Impacts on Coastal Areas", and "Climate Impacts on Ecosystems", and all links under the heading "Learn about Climate Change."
44. When will these webpage updates be completed, and what process is the EPA using to ensure that any changes to their content reflects the best available science on climate change and its impact on water resources? With respect to the water program, the EPA continues to offer the water sector and other interested stakeholders the best-available science describing how extreme weather events could affect the sector, as well as extensive information pertaining to how the sector can adopt countermeasures to reduce the risk of these impacts. Webpages are routinely updated with new information to ensure that the sector can use the best available information to adopt countermeasures to reduce the risk of these impacts.

THE HONORABLE PAUL D. TONKO

What Steps are Being Taken to Ensure the Highest Level of Adherence to EPA's Scientific Integrity Policy?

Science is the backbone of the EPA's rulemaking process. The agency's ability to pursue its mission to protect human health and the environment depends upon the integrity of the science on which it relies.

Scientific integrity is the adherence to professional values and practices when conducting, communicating, supervising, influencing and utilizing the results of scientific research. It ensures objectivity, clarity, reproducibility, and utility, while protecting against bias, fabrication, falsification, plagiarism, outside interference, and censorship. The EPA Scientific Integrity Policy is a roadmap to ensuring high standards of scientific integrity at the EPA. The policy details the components of a culture of scientific integrity, and provides a framework for agency-wide compliance. The Policy applies to all EPA employees including scientists, managers, and political appointees, as well as contractors, grantees, collaborators, and student volunteers.

The Scientific Integrity Policy also established the agency's Scientific Integrity Committee to provide oversight for its implementation. The Committee, led by the Scientific Integrity Official, encourages consistent Policy implementation, with members acting as liaisons for their offices and Regions and addressing questions and concerns regarding the Policy.

The EPA has developed a series of trainings to ensure that its employees are aware of their responsibilities under the Policy. For example, in fiscal year 2016, the EPA deployed a training program focused on increasing the awareness and understanding of the Policy and demonstrating how scientific integrity enhances the agency's work. The training was intended for employees who spend at least 25% of their time conducting, utilizing, communicating or supervising science and reached almost 6,000 EPA employees.

The EPA also is developing guidance materials to encourage a culture of scientific integrity at the agency. For example, EPA published "Best Practices for Designating Authorship" in 2016 to provide information for EPA employees, contractors, and grantees on who should be included as an author in any scientific product. Authorship is an important part of scientific integrity, as it provides transparency into the origins of a scientific product. Without knowing who was involved in the product, it is difficult to validate the merit of the work.

How are New EPA Employees, Including Political Appointees, Being Educated on These Policies? Are They Being Made Aware of What would Constitute a Violation?

Since January 2017, all new EPA employees have been required to take online scientific integrity training. The training consists of a video showing the Scientific Integrity Official conducting a training session featuring an introductory whiteboard video and discussion, followed by a short quiz. The training also includes information about what would constitute a violation of the Policy. The training helps new employees establish a personal commitment to scientific integrity, which contributes to the overall culture of scientific integrity at the EPA.

The EPA also holds an annual "Employee Conversation with the Scientific Integrity Official." This conversation serves as the annual update regarding scientific integrity at the EPA for all employees. The Scientific Integrity Official uses this opportunity to highlight the importance of scientific integrity, to discuss new initiatives, and to answer any questions.

Throughout the year, the Scientific Integrity Official provides outreach on scientific integrity, including presentations at EPA program offices, regional offices, and laboratories; development of outreach materials to distribute across the agency; participation in conferences and other events; and hosting stakeholder meetings. The EPA also publishes an Annual Report on Scientific Integrity.

Has EPA's Scientific Integrity Official Met with or Requested a Meeting with Administrator Pruitt to Discuss EPA' s Scientific Integrity Policy and Related Procedures? If Yes, When?

Yes, the EPA's Scientific Integrity Official requested a meeting with Administrator Pruitt and his advisors to provide a briefing on the Scientific Integrity Policy and related procedures. A meeting regarding the Scientific Integrity Policy and procedures occurred in May with the Administrator's Chief of Staff, the Acting Deputy Administrator, and the Director of the Office of the Science Advisor.

What is the Role of EPA' s Advisory Committees for Ensuring Integrity of Science at the Agency?

The EPA Scientific Integrity Policy provides a framework intended to ensure scientific integrity throughout the EPA and promote scientific and ethical standards, including the use of peer review and advisory committees.

The Scientific Integrity Policy states that:

> Federal Advisory Committees are an important tool within the EPA for ensuring the credibility and quality of Agency science, enhancing the transparency of the peer review process, and providing for input from the EPA's diverse customers, partners, and stakeholders. In almost all cases, FACs meet and deliberate in public and materials prepared by or for the FAC are available to the public. Consistent with the requirements of the Federal Advisory Committee Act (5 USC Appendix 2), implementing regulations from the General Services Administration (41 CFR Part 102-3), and guidance that lobbyists not serve on FACs, the EPA's scientific or technical FACs are expected to adhere to the following procedures:
>
> - Transparent recruitment of new FAC members should be conducted through broad-based vacancy announcements, including publication in the Federal Register, with an invitation for the public to recommend individuals for consideration and submit self-nominations.
> - Professional biographical information (including current and past professional affiliations) for appointed committee members should be made widely available to the public (e.g., via a website). Such information should clearly illustrate an individual's qualifications for serving on the committee.
> - The selection of members to serve on a scientific or technical FAC should be based on expertise, knowledge, contribution to the relevant subject area, balance of the scientific

or technical points of view represented by the members, and the consideration of conflicts of interest. Members of scientific and technical FACs should be appointed as special government employees. The Agency is to make all Conflict of Interest Waivers granted to committee members publicly available (e.g., via a website).

- All reports, recommendations, and products developed by FACs are to be treated as solely the findings of such committees rather than of the EPA, and thus are not subject to Agency revision.

The Agency adheres to the current standards governing conflict of interest as defined in statutes and implementing regulations. The Office of General Counsel's Ethics Office develops standard procedures and ethics training for Special Government Employees (SGEs) who serve on scientific FACs. These procedures include the submission and review of Confidential Financial Disclosure Forms for SGEs serving on advisory committees, EPA Ethics Advisory 08-02: "Ethics Obligations for Special Government Employees", and completion of an online and/or in-person Office of Government Ethics course. Some FACs at the EPA are staffed with representative members. These committee members represent the point of view of a group or organization and are not subject to the conflict of interest requirements referenced above.

Is EPA Seeing Any Signs that the Recent Dismissal of Nine Members of the Board of Scientific Counselors will have a Larger Effect on the Membership of Other Advisory Boards?

No, we have not seen any signs that the non-renewal of members whose terms expired will have an effect on the membership of other advisory boards.

How Many Resignations have There been Related to These Dismissals?

Three.

Have Any Board Members Expressed Concerns to EPA over the Handling of These Dismissals? If Yes, What are the Details of These Concerns?

BOSC members, particularly those who resigned in response to the non-renewals, have expressed some concerns to the EPA and the press about the non-renewal of these nine BOSC members. Many of the concerns were related to the timing and perceived reasons behind the decision.

What Processes are in Place to Ensure that Any New Board Members are in Compliance with all Applicable Ethics Regulations and Free of Any Conflicts of Interest or Appearances of Being Unable to Provide Impartial Advice?

All BOSC members are appointed as Special Government Employees (SGE) as defined by 18 U.S.C. § 202. As such, they are subject to federal conflict of interest statutes codified in Title 18 of the United States Code as well as the Standards of Ethical Conduct for Employees of the Executive Branch at 5 C.F.R. Part 2635. Under these federal ethics laws and regulations, they are prohibited from carrying out their duties if they have a financial conflict of interest or an appearance of a loss of impartiality. In addition, the BOSC members must comply with financial disclosure reporting requirements and annual training requirements.

As noted in the Federal Register Notice, the EPA's evaluation of an absence of financial conflicts of interest will include a review of the "Confidential Financial Disclosure Form for Special Government Employees Serving on Federal Advisory Committees at the U.S. Environmental Protection Agency" (EPA Form 3110-48). This confidential form allows government ethics officials, who are trained career employees, to determine whether a prospective or actual BOSC member has a statutory conflict between that person's public responsibilities (which includes membership on an EPA Federal Advisory Committee) and private interests and activities, or an appearance of a loss of impartiality, as defined by Federal regulation. BOSC nominees will be evaluated based on the same criteria as nominees under the previous administration.

It is My Understanding that There are 7 Members of the EPA Science Advisory Board Whose First Terms are Ending on September 30, 2017. Will These Members be Renewed?

At this time, EPA leadership has not decided whether or not these members will be renewed.

How does EPA Define Conflict of Interest?

As an executive branch agency, the EPA's employees are subject to 18 U.S.C. § 208, the financial conflict of interest statute. Under this statute, employees are prohibited from participating personally and substantially in an official capacity in any particular matter that will have a direct and substantial affect upon his own interests or anyone imputed to him. The implementing regulations for this statute are found at 5 C.F.R. Part 2635, Subpart D and Part 2640. The definition for a disqualifying financial interest is found at 5 C.F.R. § 2635.402, "disqualifying financial interest."

Who is Responsible for Determining Whether EPA Political Appointees, Including the Administrator, have Conflicts of Interest on Certain Issues?

For Presidentially Appointed Senate confirmed (PAS) positions, including the Administrator, the Office of Government Ethics approves the ethics agreement prepared by the EPA for each nominee prior to the confirmation process. Those agreements set forth the steps that the PAS appointee will take to comply with federal ethics laws and regulations, including conflicts of interest, if confirmed. The Designated Agency Ethics Official (DAEO) and Alternate Designated Agency Ethics Official (ADAEO) are responsible for overseeing the EPA's ethics program and interpreting federal ethics laws and regulations for all EPA employees. Both of these positions are located in the Office of General Counsel.

How will EPA Ensure that Key Technical Positions at the Agency are Filled with Qualified Scientists Free from Conflicts of Interest?

All EPA employees are subject to the Standards of Ethical Conduct for Employees of the Executive Branch, 5 C.F.R. Part 2635, and the federal conflicts of interest statutes codified at Title 18 of the United States Code. In additional, political appointees are subject to Executive Order 13770 and the Trump ethics pledge that they must sign. The EPA's Office of General Counsel assists employees in understanding their ethics obligations, including financial conflicts of interest.

What Role does Independent Science have in Informing EPA Decisions to Protect Public Health and the Environment?

Environmental policies, decisions, and emergency response must be grounded, at a most fundamental level, in high-quality, objective, transparent science. This includes science conducted by the EPA, other federal agencies, industry, academia, and others.

What Kinds of Communications were Involved between the White House, Industry Organizations, and EPA Regarding Chlorpyrifos?

The EPA's Office of Pesticide Programs engages with all interested stakeholders throughout its review processes and honors pertinent meeting requests from its stakeholders. There have been a number of public comment periods on aspects of the chlorpyrifos review where the public and stakeholders are able to review the agency's documents and submit their comments for consideration in our decision-making process. Comments in response to recent requests for public comment regarding chlorpyrifos have been submitted by members of the public, federal regulatory partners, non-governmental organizations, university faculty, as well as industry.

The EPA has kept its federal regulatory partners apprised of the status of chlorpyrifos with in-person meetings and phone calls, including the Department of Agriculture and the Food and Drug Administration. The agency has responded in-kind to similar requests from technical registrants, stakeholder associations, and non-governmental organizations.

Were Scientists from the Office of Chemical Safety and Pollution Prevention, or Other Relevant EPA offices, Consulted Before Administrator Pruitt Decided not to Ban Chlorpyrifos?

Senior agency leadership received briefings on chlorpyrifos from OCSPP scientists and senior management and also received input from other relevant EPA offices before issuing the March 29, 2017, Order denying the citizen petition regarding chlorpyrifos.

What Steps has EPA Taken to Implement President Trump's Executive Order on Reducing Regulation and Controlling Regulatory Costs?

Consistent with Executive Orders 13771 (Reducing Regulation and Controlling Regulatory Costs) and 13777 (Enforcing the Regulatory Reform Agenda), the EPA has been taking a hard look at EPA regulations and the EPA's Regulatory Reform Task Force's evaluation will help identify regulations that may be appropriate for repeal, replacement, or modification. The EPA has also initiated the delay or reconsideration of multiple regulations finalized by the previous administration that may further EO 13771 implementation. As a note, the Administrative Procedure Act and other applicable laws apply to any repeal, replacement, or modification of any existing regulation that the EPA undertakes.

How is EPA Choosing Which Two Regulations to Repeal for Every New Regulation Promulgated?

The EPA is still developing our internal process to fully implement EO 13771.

Please Provide an Average Annual Cost Estimate for EPA to Run Its Energy Star Program

From FY 2007 through FY 2016, the average annual budget for the EPA to implement the ENERGY STAR program has been $48 million, which includes both staffing and contracting costs.

Since 1992, How Much have Consumers Saved in Their Utility Bills Due to Energy Star Products?

Since 1992, the ENERGY STAR program, together with its partners which currently number more than 16,000, have delivered net energy bill savings exceeding $400 billion. More than $200 billion of these net energy bill savings resulted from ENERGY STAR-certified products and homes. The rest of the savings were delivered by the ENERGY STAR Commercial Buildings and Industrial Programs.

Since 1992, How Many Tons of Greenhouse Gas Emissions have been Reduced Due to Energy Star Products?

Since 1992, the ENERGY STAR Program has helped achieve broad emission reductions, including over 2.5 billion metric tons of greenhouse gas emissions. More than 1 billion metric tons of these reductions resulted from ENERGY STAR-certified products and homes. The rest of the reductions were delivered by the ENERGY STAR Commercial Buildings and Industrial Programs.

THE HONORABLE TONY CARDENAS

1. How will the sudden removal of members of the Board of Scientific Counselors affect the research into lead in drinking water and other such research used to develop national standards to ensure our public health? The Board of Scientific Counselors (BOSC) provides the EPA with access to independent advice from non-EPA experts who are nationally renowned in their disciplines, and it does so in a transparent manner with opportunities for public input through advance review of meeting agendas, meeting documents, and charge questions. These experts provide advice, information, and recommendations to the EPA on their science and research to ensure it provides the strong, scientific foundation that informs the agency's work to protect human health and the environment. Once the vacancies on the BOSC committees are filled with scientists who have the appropriate expertise, the review of the EPA's research will resume. However, the EPA's research will continue in the interim.
2. How will the Administration's budget cuts and staffing shortages affect the EPA's ability to carry out its duties required by statute, such as its programs to ensure safe drinking water? The agency's FY 2018 budget lays out a comprehensive back-to-basics and foundational strategy to maintain core environmental protection with respect to statutory and regulatory obligations. This budget provides the direction and resources to return the EPA to its core mission of protecting human health and the environment. This can be accomplished by engaging with state, local, and tribal partners to create and implement sensible regulations that also work to enhance economic growth.
3. How will the budget cuts and staffing shortages affect the oversight and testing of water systems? The EPA will continue to partner with states, drinking water utilities, and other stakeholders to identify and address current and potential sources of drinking water contamination.
4. How will budget cuts affect the Drinking Water State Revolving Fund? The EPA's budget supports the President's focus on the nation's infrastructure. The infrastructure needs of our communities include making improvements to drinking water systems, as

well as cleaning up contaminated land. A priority for the agency is modernizing the outdated water infrastructure on which the American public depends. While most small systems consistently provide safe and reliable drinking water, many small systems face challenges with aging infrastructure, increasing costs and decreasing rates bases. Funding levels in the FY 2018 budget support the President's commitment to infrastructure repair and replacement and would allow states, municipalities, and private entities to finance high-priority infrastructure investments. The FY 2018 budget includes $863 million for the Drinking Water State Revolving Fund and $20 million for the Water Infrastructure Finance and Innovation Act (WIFIA) program.

May 18, 2017

The Honorable John Shimkus
Chairman, Environment Subcommittee
Energy and Commerce Committee
U.S. House of Representatives
Washington, D.C. 20515
The Honorable Paul Tonko
Ranking Member, Environment Subcommittee
Energy and Commerce Committee
U.S. House of Representatives
Washington, D.C. 20515

Dear Chairman Shimkus and Ranking Member Tonko,

On behalf of the members of the United Steelworkers (USW), we applaud the Environment Subcommittee for the hearing during infrastructure week on water infrastructure. Our union is very committed to the repair, rehabilitation, and rebuilding of America's water infrastructure, especially those authorized by the Safe Drinking Water Act. We are grateful for the opportunity to submit these remarks for the hearing record, and we look forward to working with the committee to achieve a legislative solution that results in a well-funded and stable reauthorization of these programs, as well as maximizes their ability to provide not only desperately-needed infrastructure improvements, but thousands of good jobs as well.

The Steelworkers union has a unique interaction with drinking water funding as our union members are involved in all aspects of our country's drinking water infrastructure.

With over 50,000 members in the iron and steel industry, our members forge and manufacture the pipes, fittings, and other materials used in many communities to transport and deliver water to Americans. USW also has a large and growing public sector, which includes close to 2,000 members who conduct work related to municipal water and wastewater treatment plants.

USW members are employed by numerous manufacturers related to water infrastructure including McWane Ductile, Mueller Water Products and Ligon Industries, to name a few. The domestic manufacturing base related to water infrastructure has the opportunity to expand. Congress must address the critical maintenance needs in drinking water infrastructure in a manner that maximizes job creating potential.

The United States uses 42 billion gallons of water a day to support daily life from cooking and bathing in homes to use in factories and offices across the country. Drinking water is delivered via one million miles of pipes across the country. Every day, nearly six billion gallons of treated drinking water are lost due to leaking pipes, with an estimated 240,000 water main breaks occurring each year.

The deferred maintenance in our country's water infrastructure creates substantial waste and inefficiencies. It is estimated that leaky, aging pipes are wasting 14 to 18% of each day's treated water; the amount of clean drinking water lost every day could support 15 million households.[5] A Chicago State University study showed that by reducing the amount of water leaked annually in the U.S. by only 5 percent would result in saving enough energy to power 31,000 homes for a year and cut 225,000 metric tons of carbon dioxide emissions.[6] Upgrading and repairing our nation's infrastructure has the potential to create hundreds of thousands of jobs. The American Society of Civil Engineers currently grades our drinking water infrastructure at a "D" level in their most recent report card.[7] Raising our drinking and clean water systems to a "B" grade over the next 10 years could support or create an estimated 144,000 jobs across the U.S. economy.[8] However it is critical for Congress to implement policies which maximize the job creation potential of increasing domestic investment in drinking water infrastructure.

USW has long advocated for domestic preferences such as "Buy America" policies to ensure tax payer dollars create American jobs. In particular, USW strongly supports the American Iron and Steel Buy America preference which has been applied to the Drinking Water State Revolving Fund since 2014. This policy has been a huge success in driving job and production growth in the waterworks manufacturing sector. For example, our employers report operating capacity at plants rising from around 25 percent to closer to 70 percent after the inclusion of Buy America requirements in water infrastructure. That is the difference between one shift of work and three shifts at a mill. This additional capacity also

---

[5] https:// www.infrastructurereportcard.org/ wp-content/ uploads/ 2017/ 01/ Drinking-Water-Final.pdf.
[6] https:// www.bluegreenalliance.org/ wp-content/ uploads/ 2016/ 08/ 102414-Making-the-Grade_vFINAL.pdf
[7] https:// www.infrastructurereportcard.org/ wp-content/ uploads/ 2017/ 01/ Drinking-Water-Final.pdf
[8] https:// www.bluegreenalliance.org/ wp-content/ uploads/ 2016/ 08/ 102414-Making-the-Grade_vFINAL.pdf

lowers the overall cost of producing pipe as plants run more efficiently. Our members are economically more secure and are more competitively manufacturing the products necessary to fulfill the needs of America's water infrastructure.

The AIS preference for the Drinking Water SRF has been annually renewed via the appropriations process, including in the just-passed FY2017 omnibus appropriations measure. By contrast, the companion program to the Drinking Water SRF, the Clean Water State Revolving Fund, has a permanent statutory application of the AIS preference, which was added in the Clean Water Act Amendments in the 2014 Water Resources Reform and Development Act. USW urges this subcommittee, as well as the full Energy and Commerce Committee, to take a similar path and include a permanent statutory application of the AIS preference for the Drinking Water SRF in any reauthorization or amendment of the Safe Drinking Water Act.

We are pleased to see that the LIFT America Act, introduced earlier this week and cosponsored by several members of the subcommittee, includes the permanent statutory application of AIS in its water title, as it did in the AQUA Act from the previous Congress. We are also pleased that the level of authorized funding for the Drinking Water SRF ($22.56 billion over 5 years) is the sort of serious investment to address a serious problem that we need. As the subcommittee considers this discussion draft and the LIFT America Act, we hope that the final package includes both adequate funding and permanent statutory application of AIS for the Drinking Water SRF. Done right, we can repair and rehabilitate America's water infrastructure and create thousands of good manufacturing jobs at the same time.

Our nation's water infrastructure is in dire need of repair to prevent leakage and protect the safety of the American public. We urge Congress to ensure that legislation to address these problems also creates and maintain good jobs for American workers. We look forward to continuing to work with the committee on this issue in the future.

Sincerely,

Holly R. Hart
Assistant to the President
Legislative Director

Tony Cárdenas
Congress of the United States
29th District, California

May 19, 2017

Chairman John Shimkus
Committee on Energy and Commerce
Subcommittee on the Environment
2125 Rayburn House Office Building
Washington, D.C. 20515

Ranking Member Paul Tonka
Committee on Energy and Commerce
Subcommittee on the Environment
2125 Rayburn House Office Building
Washington, D.C. 20515

Dear Chairman Shimkus and Ranking Member Tonka:

Thank you for holding a hearing on clean water, an issue that touches the lives of all Americans. This is an issue which we can all agree is of great importance. Moving forward, we should work together to draft and improve legislation that addresses the public health need to supply clean drinking water. We also need to boost water infrastructure projects to dismantle structures that feed toxins into our drinking supply and replace them with sustainable, modern infrastructure.

I would like to highlight that while clean water is essential for the health of all Americans, it is crucial for our most vulnerable-our children. As we know from the Flint disaster, we continue to see unsafe levels of lead in drinking water. According to a recent report in Reuters, in Los Angeles, more than 17 percent of small children tested had elevated levels of lead in their blood, far exceeding the 5 percent rate of children tested in Flint, Michigan. Lead poisoning can produce serious health and behavioral issues, particularly in young children. No family should fear that their children are ingesting elevated levels of lead, only one of various contaminants that federal, state and local agencies, as well as the water industry have to contend with. We need thoughtful legislation that ensures that states have the ability to safely and effectively manage their water supply. The issue is one of resources-more and more states are unable to compensate for the federal funding gap.

While we work on legislation to improve our water supply, I have grave concerns about the future of the Environmental Protection Agency and the current safeguards it is charged with establishing, overseeing, and enforcing.

My questions are directed at the EPA:

1. How will the sudden removal of members of the Board of Scientific Counselors affect the research into lead in drinking water and other such research used to develop national standards to ensure our public health?

2. How will the Administration's budget cuts and staffing shortages affect the EPA's ability to carry out its duties required by statute, such as its programs to ensure safe drinking water?
3. How will the budget cuts and staffing shortages affect the oversight and testing of water systems?
4. How will budget cuts affect the Drinking Water State Revolving Fund?

I look forward to hearing the EPA's answers to these questions and hope that Administrator Pruitt will fully meet EPA's statutory responsibilities to the American people.

To my colleagues, I hope that we can sit at the table together and work on bipartisan legislation so we can ensure clean drinking water for every American family.

Sincerely,

Tony Cárdenas

TONY CÁRDENAS
Member of Congress

Congress of the United States
House of Representatives
COMMITTEE ON ENERGY AND COMMERCE
2125 RAYBURN HOUSE OFFICE BUILDING
WASHINGTON, DC 20515-6115
Majority (202) 225-2927
Minority (202) 225-3641

June 9, 2017

Mr. Martin Kropelnicki
President and CEO
California Water Service Group

Dear Mr. Kropelnicki:

Thank you for appearing before the Subcommittee on Environment on Friday, May 19, 2017, to testify at the hearing entitled "H.R._, Drinking Water System Improvement Act and Related Issues of Funding, Management, and Compliance Assistance under the Safe Drinking Water Act."

Pursuant to the Rules of the Committee on Energy and Commerce, the hearing record remains open for ten business days to permit Members to submit additional questions for the record, which are attached. The format of your responses to these questions should be as follows: (1) the name of the Member whose question you are addressing; (2) the complete text of the question you are addressing in bold, and (3) your answer to that question in plain text.

To facilitate the printing of the hearing record, please respond to these questions with a transmittal letter by the close of business on Friday, June 23, 2017. Your responses should be mailed to Elena Brennan, Legislative Clerk, Committee on Energy and Commerce, 2125 Rayburn House Office Building, Washington, DC 20515.

Thank you again for your time and effort preparing and delivering testimony before the Subcommittee.

Sincerely,

John Shimkus
Chairman
Subcommittee on Environment

cc: The Honorable Paul Tonko, Ranking Member, Subcommittee on Environment
Attachment

June 16, 2017

The Honorable John Shimkus
Chair, Subcommittee on Environment

The Honorable Paul Tonko
Ranking Member, Subcommittee on Environment

c/o Elena Brennan
Legislative Clerk, Committee on Energy and Commerce
2125 Rayburn House Office Building
Washington, DC 20515

Re: Questions for the Record following Hearing on Nation's Drinking Water Infrastructure

Dear Chair Shimkus and Ranking Member Tonko:

Thank you for inviting me and the National Association of Water Companies (NAWC) to testify before the Subcommittee on Environment during its May 19, 2017 hearing on

H.R. __, Drinking Water System Improvement Act and Related Issues of Funding, Management, and Compliance Assistance Under the Safe Drinking Water Act.

Again, I commend you and the Subcommittee for highlighting the challenges facing the country's drinking water systems and the solutions that will help ensure all Americans have safe, reliable, and high-quality water utility service for generations to come. California Water Service (Cal Water) and NAWC's other member companies stand ready, able, and willing to work with all levels of government to help overcome these challenges.

Enclosed you will find NAWC's responses to the additional questions for the record you submitted. If you have any questions, please do not hesitate to reach out to me. We look forward to continuing to work with the Subcommittee and Congress on the critical issues associated with the nation's water infrastructure.

Sincerely,

Martin A. Kropelnicki

Martin A. Kropelnicki
President & CEO

Enclosure
Quality. Service. Value.
calwater.com

Answers to Questions for the Record Following a Hearing Entitled "H.R., Drinking Water System Improvement Act and Related Issues of Funding, Management, and Compliance Assistance under the Safe Drinking Water Act" Conducted by the Subcommittee on Environment, House Energy and Commerce Committee

On May 19, 2017, the Subcommittee on Environment of the House Committee on Energy and Commerce convened a hearing entitled "H.R., Drinking Water System Improvement Act and Related Issues of Funding, Management, and Compliance Assistance under the Safe Drinking Water Act," at which Martin A. Kropelnicki, President & CEO of California Water Service Group and President of the National Association of Water Companies (NAWC), testified on behalf of NAWC about the ways the private water sector can help address the nation's drinking water infrastructure challenges. Chairman Shimkus submitted further questions for the record, and this document provides NAWC's responses.

THE HONORABLE JOHN SHIMKUS, CHAIRMAN, SUBCOMMITTEE ON ENVIRONMENT

Question: Many of the groups on this panel were part of EPA's report on Effective Utility Management. Do you have specific recommendations on what Congress should do, if anything, with that report?

Congress can take several steps that will help further the implementation of Effective Utility Management (EUM) and, at the same time, help address the significant drinking water infrastructure challenges the country is facing. First, as a general rule, applicants for public funding of drinking water projects should demonstrate that they have fully accounted for the long-term costs of their projects, including any risks inherent in construction, operations, and/or maintenance, and have selected the delivery model that provides the best long-term value to the water supplier's customers. For a community to maintain and improve the condition of its infrastructure, and to ensure its long-term safety and reliability, water utilities should be expected, at a minimum, to manage their assets based on a process where adequate repair, rehabilitation, and replacement are fully reflected in management decisions and fully accounted for in water rates. Failing water systems should not be subsidized without an expectation of financial and operational viability and a process to ensure that federal funds are targeted in a way to ensure they are being used efficiently and cost-effectively.

Second, especially in situations where water suppliers are unable or unwilling to operate their systems in accordance with the principles of EUM, Congress could take steps to further prioritize and incentivize partnerships between failing water systems and owners or operators that have a strong track record of providing safe, reliable, and high-quality service to their customers. For example, Congress should establish a more robust legal "safe harbor" for water suppliers that assume the responsibility of owning and/or operating failing and noncompliant water systems. Oftentimes, the legal and financial liabilities of distressed systems, which can range from the hundreds of thousands to millions of dollars, serve as a "poison pill" to prospective operators or owners. A more robust legal "safe harbor" would prevent new operators or owners from being held liable for the previous misdeeds of others and, in the process, open the doors to significant amounts of capital being invested into the nation's drinking water infrastructure.

Finally, in order to increase the level of private investment in our drinking water systems, Congress should explore the possibility of creating a tax-based incentive for private water companies that enter into consolidation or partnership arrangements with noncompliant systems. In those cases where the noncompliant system is publicly owned, the federal government is already not receiving any income tax revenue from the water system. It may make sense to extend that income tax benefit to a private water company that assumes responsibility for the noncompliant system, either for a certain number of years or until the failing system is brought into compliance. In the short-term, such an incentive would be revenue neutral, and over the medium-and long-term, it would be a revenue enhancer. In addition to creating an incentive for more partnerships and consolidations, this approach would help to address some short-term affordability questions and free up additional capital to be invested into the water systems.

In summary, what is needed to address the nation's drinking water infrastructure challenges is a willingness to explore innovative solutions such as partnerships and incentivized consolidation. While many communities continue to clamor for more federal funding, more funding is not going to solve this growing crisis. In many cases, water system failures – be they related to water quality, reliability, or both – are not solely due to the absence of funding, but rather are directly attributable to the failure of proper governance, poor decision-making, and lack of stringent oversight.

Question: If Congress should do nothing, what should utilities do to facilitate action?

One of the most important steps any utility can take to help ensure that it is able to provide its customers with safe, reliable, and high-quality service is to manage its assets in such a way that adequate repair, rehabilitation, and replacement are fully reflected in management decisions, including water pricing. In 2003, the EPA established its Four Pillars of Sustainable Infrastructure, one of which was full-cost pricing. Nearly a decade and a half later, thousands – if not tens of thousands – of water utilities across the country have water utility rates that do not reflect the actual cost of operating, maintaining, and upgrading their systems.

Quite simply, full-cost pricing of water utility service is the single most important element of any strategy to improve the nation's drinking water infrastructure and compliance with the country's water quality standards. Full-cost pricing helps to ensure the financial viability of water suppliers, which then enables the supplier to undertake needed maintenance of and upgrades to its facilities, both of which play a critical role in the supplier's ability to provide safe and high-quality water to its customers.

This transition to full-cost pricing should, however, be accompanied by adequate financial support to assist economically distressed communities and low-income households. In this regard, Congress may wish to consider providing relief directly to fixed- and low-income customers. Currently, federal funds flow directly to water utilities, which enable them to charge lower rates to all of their customers, including those who are not facing any type of economic hardship. A more efficient approach may be to transfer funds directly to challenged and low-income customers, similar to the Low Income Home Energy Assistance Program for gas and electric customers.

Congress of the United States
House of Representatives
COMMITTEE ON ENERGY AND COMMERCE
2125 RAYBURN HOUSE OFFICE BUILDING
WASHINGTON, DC 20515-6115
Majority (202) 225-2927
Minority (202) 225-3641

June 9, 2017

Mr. Scott Potter
Director
Nashville Metro Water Services

Dear Mr. Potter:

Thank you for appearing before the Subcommittee on Environment on Friday, May 19, 2017, to testify at the hearing entitled "H.R._, Drinking Water System Improvement Act and Related Issues of Funding, Management, and Compliance Assistance under the Safe Drinking Water Act."

Pursuant to the Rules of the Committee on Energy and Commerce, the hearing record remains open for ten business days to permit Members to submit additional questions for the record, which are attached. The format of your responses to these questions should be as follows: (1) the name of the Member whose question you are addressing, (2) the complete text of the question you are addressing in bold, and (3) your answer to that question in plain text.

To facilitate the printing of the hearing record, please respond to these questions with a transmittal letter by the close of business on Friday, June 23, 2017. Your responses should be mailed to Elena Brennan, Legislative Clerk, Committee on Energy and Commerce, 2125 Rayburn House Office Building, Washington, DC 20515.

Thank you again for your time and effort preparing and delivering testimony before the Subcomittee.

Sincerely,

John Shimkus
Chairman
Subcommittee on Environment

cc: The Honorable Paul Tonka, Ranking Member, Subcommittee on Environment
Attachment

Response to Questions Submitted to:
Scott Potter
Director

Nashville Metro Water Services Testifying on the behalf of Association of Metropolitan Water Agencies (AMWA) Regarding Hearing on "H.R.___, Drinking Water

System Improvement Act and Related Issues of Funding, Management, and Compliance Assistance under the Safe Drinking Water Act."

May 19, 2017

THE HONORABLE JOHN SHIMKUS

1. Safe Drinking Water Act Section 1433 calls on community water systems to conduct vulnerability assessments of their systems to terrorist attack or other intentional acts designed to disrupt the ability of the water system to provide a safe and reliable supply of drinking water.
   (a) How recently has your utility reviewed and updated your vulnerability assessment? At Nashville Metro Water Services, we began a process to review and update our vulnerability assessment beginning in December 2016. We just completed this process in May 2017. This assessment included an all hazards analysis, meaning that we reviewed the system's vulnerability to not only terrorists or other intentional acts, but also natural disasters such as flooding. We plan to continue to periodically update the vulnerability assessment in the future as circumstances warrant, such as when we encounter a new operating environment or when we become aware of a new type of threat. Nashville Metro Water Services also maintains an up-to-date emergency response plan, which outlines plans and procedures for responding to threats identified in our vulnerability assessment. The utility reviews and updates its emergency response plan annually.
   (b) Is your utility unique among your peers in reviewing your vulnerability assessment without a government mandate to do so? No, Nashville Metro Water Services is not unique in this respect. In fact, based on discussions I have had with managers of other large utilities that are members of professional organizations such as AMWA and the American Water Works Association (AWWA), it is a common practice for these drinking water systems to update their vulnerability assessments and emergency response plans without being mandated to do so. Because large water systems have made it a practice to keep their vulnerability assessments and emergency response plans up-to-date, Congress must keep these systems in mind in the event that it considers legislation to require these documents to be updated by a certain date. For example, any new law that mandates an update of vulnerability assessments or emergency response plans should include a "grandfather clause" that exempts utilities from having to immediately redo these assessments again if they certify that they had already reviewed and updated the documents within the previous two years.

2. I appreciate the forthrightness of your testimony when it comes to suggesting a guideline for what your organization believes is the correct number to fund the Drinking Water State Revolving Loan Fund. Does it matter to you whether the number is flat each fiscal year – meaning it would be the same each year – or having it steadily increase every year? EPA's Drinking Water Needs Surveys, completed ever four years, have consistently found that communities' drinking water infrastructure spending needs will grow in the years and decades ahead. As such, AMWA believes it is appropriate for the Drinking Water SRF's authorized funding level to increase each year as well. As my written testimony explains, a DWSRF authorization level of $1.8 billion is a reasonable starting point because it is roughly double the program's most recent annual appropriation and would not immediately constrain the ability of Congress to deliver adequate funding to the program. While Congress must remain cognizant of states' financial ability to meet their 20 percent funding match, looking ahead the committee should consider increasing the authorization each year at least until it reaches about $2.7 billion, a sum that aligns with President Trump's previous call to triple DWSRF funding. Finally, I should note that when Congress authorized the Water Infrastructure Finance and Innovation Act (WIFIA) pilot program in 2014, it chose to increase the program's authorization by 250 percent over five years. So there is ample precedent for Congress to steadily increase the authorization level of a program to aid local water infrastructure financing efforts.
3. Your colleague, Rudy Chow from Baltimore, MD, in a written response to a question from our last drinking water hearing, mentioned that codifying the EPA's current practice for Consumer Confidence Reports is among the most significant, non-financial areas where Congress can assist drinking water systems. Can you explain that point for me? As a result of a regulatory review carried out under President Obama, EPA revised its interpretation of the Safe Drinking Water Act's requirement that community water systems deliver their customers a copy of a consumer confidence report each year. Under EPA's new interpretation, community water systems were given the option to deliver these reports to customers electronically, such as by posting the reports publicly online and notifying customers of their availability via notices on water bills. Conversely, water systems that prefer to deliver hard copies of these reports to their customers may continue to do so, as they always have. The new flexibility offered by EPA's policy has brought significant savings to water systems and their ratepayers nationwide. For example, as a result of the new policy 2012 was the last year that Nashville Metro Water Services printed and mailed the full Consumer Confidence Report to all customers. That year, we mailed 155,488 individual copies of the CCR, with total printing, handling and postage costs totaling $42,631. Since 2013 we have posted the full CCR online and mailed a reminder postcard to all of our customers with a direct URL and instructions for accessing it. As a result, our per-unit cost for mailed CCR communications has decreased compared to five years ago, in

spite of higher costs for postage and supplies. Many other utilities across the country have realized even greater savings by including the notice about CCR availability on or alongside billing statements that are sent to customers. Nashville's experience appears to be typical of many other metropolitan water systems. For example, a 2016 survey of AMWA members found that 80 percent of responding utilities used electronic CCR delivery last year. These utilities reported avoiding printing an average of more than 138,000 paper CCRs, and saved an average of $44,205 in printing and postage costs. Assuming that these figures are representative of all community water systems in the U.S. that serve more than 100,000 people, fully adopting electronic CCR delivery nationwide would save more than 55 million pieces of paper and nearly $17.7 million just at the country's 400 largest water systems. These savings represent additional resources that communities are able to devote to infrastructure investment. AMWA supports Congress taking the opportunity of a DWSRF reauthorization bill to codify this EPA policy in the SDWA statute, thus ensuring that the ability to utilize electronic delivery options may not be unilaterally removed by a future EPA administrator.

4. Your testimony mentions that there are places in the Safe Drinking Water Act and the Drinking Water State Revolving Loan Fund program that do not need "top-to-bottom overhaul." So that Congress does no harm, outside of mandatory deadlines and the contaminant regulatory process which you already mentioned, can you give me examples of areas you think would not need the "top-to-bottom overhaul"? AMWA is aware of proposals that would require public water systems to assess potential threats related to climate change and nearby industrial and agricultural activities. Water utilities would have to repeatedly resubmit these assessments to EPA, along with documents outlining strategies to mitigate these threats, and emergency response plans detailing how the water system would respond in the event that one of these hypothetical risks played out. While I'm sure these proposals come from a good place, it would take a tremendous amount of resources for a water utility to develop a detailed plan that accounts for each possible risk related to climate change, plus an inventory of the ways the utility could mitigate this range of risks, plus an emergency response plan to guide the response should any one of these risks come to pass. Given that Nashville's most recent vulnerability assessment review and update took six months to complete, mandating even more requirements would quickly become a never-ending exercise. AMWA also does not believe Congress should legislate particular disinfectant methods or chemicals used by water systems. We believe local water utility experts are best equipped to determine the optimal disinfectant to protect public health and ensure compliance with SDWA, so no future SDWA reforms should attempt to broadly steer all utilities away from one disinfection method or another. Finally, AMWA believes Congress could maintain the integrity of SDWA's regulatory process by directing EPA to develop consistent practices to govern the future development of

health advisories. Section 1412 of SDWA allows the EPA Administrator to publish health advisories for contaminants that are not subject to any national primary drinking water regulation. Health advisories are therefore an important tool for providing information on emerging risks, particularly in regions that may have exposure to a particular contaminant that does not meet the threshold for development of a NPDWR. Health advisories are not regulations, but have the real potential to become de facto regulations given resource constraints at the Federal and State level. To avoid potential regulatory confusion, Congress should require EPA to develop criteria and an open process for the development of health advisories and to report back to Congress within the next 180 days laying out criteria and a process for how they are formulated.

THE HONORABLE RICHARD HUDSON

1. Your testimony called for allowing drinking water state revolving loan funds to be used for water system security enhancements. How often do water systems engage in vulnerability assessments or site security plans? Is that true for the other water utility members of the panel? It is a common practice for large community water systems to periodically review and update their vulnerability assessments and emergency response plans to ensure they are consistent with the current characteristics of the facility and account for known threats. In Nashville, we began our most recent vulnerability update in December 2016 and completed it in May 2017. We update our emergency response plan on an annual basis.
2. What types of items are you looking to have covered that are not otherwise covered by the Drinking Water State Revolving Loan Fund?" AMWA believes that Congress should formally allow community water systems to access Drinking Water SRF funds for security enhancements. After 9/11 EPA clarified that DWSRF dollars may be used for water facility security enhancements like fencing, security cameras and lighting, motion detectors, and redundant power systems, and EPA continues to recognize such expenses as eligible today. We are not looking to expand this eligibility, but we do believe it would be worthwhile for Congress to codify in the SDWA statute that DWSRF funds may be used for security measures. This would remove any risk of EPA revising its interpretation of the statute in the future, and would align the statutory DWSRF eligibilities with those of the CWSRF, which in 2014 were expanded by Congress to include "measures to increase the security of" treatment works.

THE HONORABLE PAUL D. TONKO

1. Systems have a hard time attracting talented and qualified employees. Many young people do not know these career opportunities exist. Meanwhile existing employees are getting closer to retirement. There is a lot of institutional knowledge at stake. Do you have any recommendations on what can be done to develop the water utility workforce? Developing a sustainable water utility workforce is one of the most pressing personnel challenges faced today by the drinking water community. In particular, drinking water utilities face strong competition from other sectors to recruit and retain skilled college graduates. Utilities should start thinking about innovative strategies to develop the water utility workforce, such as partnering with local colleges and universities to develop curriculums that could position graduates for long-term careers. Similarly, we need to reach out to stakeholders in our local communities to connect with local residents who may be able to fill some of the vital positions on the utility staff that do not require a college education. Of course, key to maintaining a strong workforce is having the ability to offer competitive pay and benefits, so employees are eager to stay with the utility for the long-term. But doing this requires adequate budget space, so it is important that we keep other manageable costs down so that we can pay our employees what they expect to earn. Maintaining access to low-cost infrastructure financing, such as through tax-exempt municipal bonds, is one way to keep the capital project side of the budget in check so that we have more resources to devote to our workforce.

Congress of the United States
House of Representatives
COMMITTEE ON ENERGY AND COMMERCE
2125 RAYBURN HOUSE OFFICE BUILDING
WASHINGTON, DC 20515–6115
Majority (202) 225-2927
Minority (202) 225-3641

June 9, 2017

Ms. Lisa Daniels
Director
Bureau of Safe Drinking Water
Pennsylvania Department of Environmental Protection

Dear Ms. Daniels:

Thank you for appearing before the Subcommittee on Environment on Friday, May 19, 2017, to testify at the hearing entitled "H.R._, Drinking Water System Improvement Act and Related Issues of Funding, Management, and Compliance Assistance under the Safe Drinking Water Act."

Pursuant to the Rules of the Committee on Energy and Commerce, the hearing record remains open for ten business days to permit Members to submit additional questions for the record, which are attached. The format of your responses to these questions should be as follows: (1) the name of the Member whose question you are addressing, (2) the complete text of the question you are addressing in bold, and (3) your answer to that question in plain text.

To facilitate the printing of the hearing record, please respond to these questions with a transmittal letter by the close of business on Friday, June 23. Your responses should be mailed to Elena Brennan, Legislative Clerk, Committee on Energy and Commerce, 2125 Rayburn House Office Building, Washington, DC 20515.

Thank you again for your time and effort preparing and delivering testimony before the Subcommittee.

Sincerely,

John Shimkus
Chairman
Subcommittee on Environment

cc: The Honorable Paul Tonko, Ranking Member, Subcommittee on Environment
Attachment

ATTACHMENT – ADDITIONAL QUESTIONS FOR THE RECORD

The Honorable John Shimkus

1. There has been some discussion about the role of using asset management as a criteria when disbursing SRF loans. In section 1452(a)(3)(A), there is a requirement prohibiting funding for a public water system which does not have the "technical, managerial, and financial capacity to ensure compliance."
   (a) What role does asset management play in compliance with this requirement of law and of SDWA section 1420?
   (b) What role does review of a utility's rates play?
2. As you mentioned in your testimony, from 1996 to 2013, the national compliance percentage with health-based standards for water systems has increased from 85% to 93%. A lot of times in Congress we only hear about the nation's problems, so it

is nice to hear this positive statistic and we of course want to see that compliance percentage continue to rise.

(a) What other positive trends or success stories are happening with our nation's drinking water infrastructure?

(b) This statistic on improved water quality compliance seems to be contrary to the fact that our nation's water infrastructure is in dire need of repair and investment. How do you explain this discrepancy?

The Honorable Debbie Dingell

1. Ms. Daniels, in Pennsylvania, how is your department working to improve communication and notification of water quality with the public?

June 19, 2017

Ms. Elena Brennan
Legislative Clerk
Committee on Energy and Commerce

Dear Ms. Brennan:

As requested, please find below my responses, as President-Elect of the Association of State Drinking Water Administrators (ASDWA), to questions posed by Chairman Shimkus and Mrs. Dingell in your letter of June 9, 2017. The questions relate to my testimony before the Subcommittee on Environment during the May 19 hearing titled "HR__, Drinking Water System Improvement Act and Related Issues of Funding, Management, and Compliance Assistance under the Safe Drinking Water Act."

Please express our appreciation to Chairman Shimkus and the Subcommittee for the opportunity to testify and provide additional information.

From the Honorable John Shimkus

1. There has been some discussion about the role of using asset management as a criteria when disbursing SRF loans. In Section 1452(a)(3)(A), there is a

requirement prohibiting funding for a public water system which does not have "the technical, managerial, and financial capacity to ensure compliance."

A. What role does asset management play in compliance with this requirement of law and of SDWA section 1420?
B. What role does review of a utility's rates play?

(1A) When designing the 1996 SDWA Amendments, Congress recognized that many drinking water utilities – and especially the smaller systems – did not have all of the elements necessary to attain and sustain their abilities to meet Federal compliance requirements. While creation of the DWSRF provided the financial wherewithal for many water utilities to achieve and maintain compliance with national primary drinking water regulations, it by no means was a silver bullet. To enhance the success of both the DWSRF and public health protection, Congress created the capacity development program (SDWA §1420) that allows states to work with struggling systems to help them achieve technical, managerial, and financial capabilities to meet Federal drinking water requirements. In the early years, states focused most of their efforts on supporting systems' technical needs – how to take samples, maintain a monitoring schedule, pass a sanitary survey. As systems gained confidence in their abilities to achieve operational proficiency, states began to look more closely at small system managerial and financial capabilities.

One of the tools that has been very helpful in educating drinking water utilities on how to attain and maintain their systems has been the development of asset management programs. For more than 10 years, under the auspices of capacity development strategies, states have been working directly and through their contracted assistance providers to educate smaller systems about the concepts and application of asset management principles as a road to successful public health protection. Many small systems are intimidated by the process and require direct, one-on-one training and support. EPA's university-based environmental finance centers have developed numerous webinars and support resources for asset management and the Agency's Office of Ground Water and Drinking Water has continued to update and improve its asset management tool (Check Up Program for Small Systems). Other, larger systems have taken advantage of the AWWA asset management tool.

Successfully gaining the confidence of small drinking water systems to undertake and sustain an asset management program is a long term effort. In the early days, many of these systems did not have a basic business plan and kept their financial records in a shoebox. Educating them on the value of instituting asset management, explaining the process, and having these systems follow through is not something that happens easily or quickly. The Capacity Development program as outlined in SDWA §1420 is an invaluable resource in helping smaller drinking water systems take on asset management which, in turn, enhances their eligibility for a DWSRF loan.

(1B) Reviewing utility rates is an activity not traditionally undertaken by state drinking water programs. Rate reviews, rate structures, and rate changes are generally managed by state Public Utilities Commissions or Public Service Commissions. Most states do not engage in local decisionmaking when it comes to rates. State drinking water programs do, however, provide outreach and education to community water systems about the different types of rate structures, the value of choosing the right rate structure, and the resources available to help a system make those determinations. States also work with smaller systems to understand asset management and how rates may affect the ability of the system to operate effectively and efficiently.

2. As you mentioned in your testimony, from 1996 to 2013, the national compliance percentage with health-based standards for water systems has increased from 85% to 93%. A lot of times in Congress we only hear about the nation's problems, so it is nice to hear this positive statistic and we of course want to see that compliance percentage continue to rise.
   A. What other positive trends or success stories are happening with our nation's drinking water infrastructure?
   B. This statistic on improved water quality compliance seems to be contrary to the fact that our nation's water infrastructure is in dire need of repair and investment.. How do you explain the discrepancy?

(2A) In general, the overall number of public water systems in the US has declined by nearly 20,000 since 1996 (170,942 v. 151,137). This is a good news story because it reflects the thoughtful consideration of many former systems as they learned about their responsibilities. Many did not even know that they had water system responsibilities. Nearly 4,100 of the reduced system number applies to community water systems...those that serve year-round populations of more than 25 people. These systems, for the most part, declined to continue as a drinking water system and were absorbed into a neighboring community system, joined forces with co-located systems to create a larger operational unit, or, in the more remote areas, simply dissolved and returned to private wells. These restructuring efforts have reduced the number of unsustainable (struggling?) systems and served to enhance our public health protection abilities as reflected in the compliance numbers referenced in the question.

Similarly, through the capacity development and operator certification elements of the 1996 SDWA, concerted education and outreach to small drinking water systems and their operators has resulted in better performing systems, better trained and educated operators, and a greater understanding of the 'why' behind many of the new rules and regulatory requirements. Source water protection efforts are another non-regulatory element in water system successes. Simply knowing your water source and taking simple steps toward prevention provide a significant reduction in the costs to remove known contaminants from the water supply and diminishes the downstream impacts of wastewater treatment. Finally, since the first infusion of $358.6 million in Federal funds (FY 97), the DWSRF has funded more than 13,000 projects for drinking water systems across the nation. Because of the availability of these funds, many of the repairs and upgrades needed to maintain system integrity were implemented and water quality and quantity problems were resolved. Cumulatively, between FY 97 and FY 16, the Federal investment in the DWSRF has been nearly $18.4 billion and states have contributed an additional $3.45 billion. In addition, many states have leveraged the core funds to provide even more money for loans to drinking water systems.

(2B) There are really two components to our response to this question. While everyone agrees that the DWSRF has been successful and provides critically needed funding to meet the infrastructure needs of the drinking water utilities across the Nation, not all public health and compliance problems are rooted in physical infrastructure. The 1996 Amendments to the SDWA offered opportunities to delineate, assess, and protect source waters; to train and educate water system operators; to help struggling systems understand their managerial and financial responsibilities; to implement new rules that offer greater public health protection; and to communicate more effectively and efficiently with the public about the quality of the water they drink. Each of these factors contribute to improved public health protection and greater compliance, yet are not directly connected to aging infrastructure.

Separately, the referenced health based statistics do not always reflect the breadth of problems found at a water system. Problems caused by aging or inadequate infrastructure often show up as significant deficiencies during sanitary surveys, where the deficiencies can be identified and addressed through the find-and-fix provisions of the RTCR, GWR, and SWTRs. In addition, aging/inadequate infrastructure often results in the complete failure of a piece of equipment or a facility, such as a breakdown in chemical feed equipment or pumps, or a failure of water mains, pipes, valves or other appurtenances. And these failures often lead to dire consequences such as water outages, boil water advisories, or “do not consume” or “do not use” notices. Here are a few recent examples from Pennsylvania:

On December 1, 2016, the Carlisle Borough Municipal Authority experienced a catastrophic failure of their water filtration plant due to an equipment failure. A check valve failed on the discharge side of a high service pump, and allowed the water from two large finished water storage tanks to flow downhill back into the filter plant at an estimated rate of more than 4,000 gallons per minute; causing the clearwell to overflow and flood the below-ground pipe gallery. Multiple pieces of equipment were submerged and destroyed or rendered non-functional, including raw water pumps and motors, high service pumps and motors, water quality monitoring equipment, and some of the chemical feed equipment. The filter plant was rendered inoperable, and Carlisle was forced to implement mandatory water use restrictions and utilize several permanent and temporary emergency interconnections with adjacent water suppliers. The mandatory restrictions were in place until December 7, when Carlisle was finally able to complete repairs and/or replacements and resume production.

Since early 2016 and continuing into 2017, the Pittsburgh Water and Sewer Authority has been cited for multiple violations and deficiencies, several of which are the direct result of aging/inadequate infrastructure. These situations have included breakdowns in chemical feed equipment, failure of a rising main, treatment efficacy issues at a membrane filtration plant, problems with several pump stations, and concerns about the integrity of their clearwell. And while these violations have not resulted in MCL exceedances, they have most definitely resulted in multiple field orders, necessary emergency corrective actions, and several boil water advisories. Work at this system is ongoing to bring them back into compliance and ensure public health protection.

In summary, the improved compliance rates, while not always tied directly to aging infrastructure, do not counter the need for infrastructure funding; rather, taken together compliance rates will continue to improve as infrastructure needs are met.

From the Honorable Debbie Dingell

3. Ms. Daniels, in Pennsylvania, how is your department working to improve communication and notification of water quality with the public?

While there is always room for improvement, Pennsylvania has worked hard over the years to ensure access to water quality information and improve transparency. Since 2002, all public water system compliance sample results have been publicly available on our website through the Drinking Water Reporting System available at http://www.drinking water.state.pa.us/dwrs/HTM/Welcome.html . And since 2009, Pennsylvania has required mandatory electronic reporting of compliance data to ensure data integrity. All data that is submitted undergoes multiple QA/QC batch edit checks, and is then run through automated compliance programs to determine MCL compliance. The public can access sample results, inventory information and violation data for all 8,500+ public water systems in the state. PA is also working on enhancements to our Department-wide enterprise system – eFACTS – to provide better access to inspection results and permitting data.

Regarding notification of water quality problems, Pennsylvania enacted more stringent public notification requirements in 2009 to improve the delivery and effectiveness of public notice for our most serious violations – Tier 1 violations. Pennsylvania also has a long-standing requirement that water suppliers must notify the Department within one hour of becoming aware of a violation or situation with the potential for adverse impacts on water quality or quantity. This allows us to immediately consult with the water supplier about the situation, and make very quick decisions about actions that may be needed to protect public health.

Areas for improvement include transitioning to electronic inspections and electronic permitting. This would allow us to make information more readily available and accessible. State resources have been a challenge to making this a reality

Sincerely,

Lisa D. Daniels
ASDWA President-Elect and
Director, Bureau of Safe Drinking Water
Pennsylvania Department of Environmental Protection

Congress of the United States
House of Representatives
COMMITTEE ON ENERGY AND COMMERCE
2125 Rayburn House Office Building
Washington, DC 20515-6115
Majority (202) 225-2927
Minority (202) 225-3641

June 9, 2017

Mr. Kurt Vause
Special Projects Director
Engineering Division
Anchorage Water and Wastewater Utility

Dear Mr. Vause:

Thank you for appearing before the Subcommittee on Environment on Friday, May 19, 2017, to testify at the hearing entitled "H.R._, Drinking Water System Improvement Act and Related Issues of Funding, Management, and Compliance Assistance under the Safe Drinking Water Act."

Pursuant to the Rules of the Committee on Energy and Commerce, the hearing record remains open for ten business days to permit Members to submit additional questions for the record, which are attached. The format of your responses to these questions should be as follows: (1) the name of the Member whose question you are addressing, (2) the complete text of the question you are addressing in bold, and (3) your answer to that question in plain text.

To facilitate the printing of the hearing record, please respond to these questions with a transmittal letter by the close of business on Friday, June 23, 2017. Your responses should be mailed to Elena Brennan, Legislative Clerk, Committee on Energy and Commerce, 2125 Rayburn House Office Building, Washington, DC 20515.

Thank you again for your time and effort preparing and delivering testimony before the Subcommittee.

Sincerely,

John Shimkus
Chairman
Subcommittee on Environment

cc: The Honorable Paul Tonko, Ranking Member, Subcommittee on Environment
Attachment

ATTACHMENT – ADDITIONAL QUESTIONS FOR THE RECORD

The Honorable John Shimkus

1. Safe Drinking Water Act Section 1433 calls on community water systems to conduct vulnerability assessments of their systems to terrorist attack or other intentional acts designed to disrupt the ability of the water system to provide a safe and reliable supply of drinking water.
   (a) Could you please explain what AWWA G430-14 and ANSIIAWWA J-100 are?
   (b) Are these standards used by AWWA members and what do they cover?
   (c) Are there other, non-mandatory standards that water utilities are using to update their vulnerabilities to various hazards?

The Honorable Paul D. Tonko

1. Systems have a hard time attracting talented and qualified employees. Many young people do not know these career opportunities exist. Meanwhile existing employees are getting closer to retirement. There is a lot of institutional knowledge at stake. Do you have any recommendations on what can be done to develop the water utility workforce?

The Honorable Debbie Dingell

1. Mr. Vause, how can federal infrastructure investment be used to modernize service lines and protect communities from lead and other contaminants?
2. Mr. Vause, does the bill we are considering today provide the funding necessary to make robust investments in our drinking water infrastructure? And what would be the impact nationwide if funding was decreased or cut?
3. Mr. Vause, what else needs to be done to address the risk of lead in drinking water – how important is it to finalize revisions to the Lead and Copper rule?

Congress of the United States
House of Representatives
COMMITTEE ON ENERGY AND COMMERCE
2125 Rayburn House Office Building
Washington, DC 20515–6115
Majority (202) 225-2927
Minority (202) 225-3641

Ms. Lynn Thorp
National Campaigns Director
Clean Water Action

Dear Ms. Thorp:

Thank you for appearing before the Subcommittee on Environment on Friday, May 19, 2017, to testify at the hearing entitled "H.R._, Drinking Water System Improvement Act and Related Issues of Funding, Management, and Compliance Assistance under the Safe Drinking Water Act."

Pursuant to the Rules of the Committee on Energy and Commerce, the hearing record remains open for ten business days to permit Members to submit additional questions for the record, which are attached. The format of your responses to these questions should be as follows: (1) the name of the Member whose question you are addressing, (2) the complete text of the question you are addressing in bold, and (3) your answer to that question in plain text.

To facilitate the printing of the hearing record, please respond to these questions with a transmittal letter by the close of business on Friday, June 23, 2017. Your responses should be mailed to Elena Brennan, Legislative Clerk, Committee on Energy and Commerce, 2125 Rayburn House Office Building, Washington, DC 20515.

Thank you again for your time and effort preparing and delivering testimony before the Subcommittee.

Sincerely,

John Shimkus
Chairman
Subcommittee on Environment

cc: The Honorable Paul Tonko, Ranking Member, Subcommittee on Environment
Attachment

June 23, 2017

The Honorable John Shimkus
U.S. House of Representatives
Committee on Energy and Commerce
Washington DC 20515-6115

Dear Representative Shimkus,

Thank for you for the opportunity to appear before the Subcommittee on Environment to testify at the hearing on the "Drinking Water System Improvement Act and Related Issues of Funding, Management, and Compliance Assistance under the Safe Drinking Water Act."

In response to the questions from the Honorable Debbie Dingell:

1. Ms. Thorp, do you think the Safe Drinking Water Act should be amended to strengthen notification requirements and increase transparency?
2. Ms. Thorp, does the draft bill under discussion today include any provisions that would improve or strengthen public notification requirements?

There are a number of ways that Safe Drinking Water Act implementation could be improved to increase transparency and to improve both the Public Notification requirements that pertain to all National Primary Drinking Water Regulations and the specific public education and notification requirements in the Lead and Copper Rule. For example, general public notification materials and requirements could be updated to reflect changes in use of traditional and social media and social science research on how people understand public health risk. In the case of lead, the U.S. Environmental Protection Agency (EPA) has identified transparency as a key element of its November 2016

"Drinking Water Act Plan," and has asked states to work with water systems to improve access from everything from monitoring data to the location of lead service lines.

While many improvements to implementation can be done through revisions to regulations and guidance, updates to the underlying statute can help to reinforce the importance of these activities. There are relevant provisions in H.R. 1068, the Safe Drinking Water Act amendments of 2017, introduced earlier this year.

The bill under discussion at the May 19 hearing, referenced above, did not include provisions related to public notification or transparency to my knowledge.

Thank you again for the opportunity to testify and please let me know if I can be of further assistance.

Sincerely,

Lynn Thorp
National Campaigns Director

In: The Drinking Water System Improvement Act
Editor: Noah L. Hogan
ISBN: 978-1-53617-170-9

Chapter 2

DRINKING WATER SYSTEM IMPROVEMENT ACT OF 2017*

Committee on Energy and Commerce

Mr. Walden, from the Committee on Energy and Commerce, submitted the following REPORT together with ADDITIONAL VIEWS [To accompany H.R. 3387] [Including cost estimate of the Congressional Budget Office]

The Committee on Energy and Commerce, to whom was referred the bill (H.R. 3387) to amend the Safe Drinking Water Act to improve public water systems and enhance compliance with such Act, and for other purposes, having considered the same, report favorably thereon with an amendment and recommend that the bill as amended do pass.

The amendment is as follows:

Strike all after the enacting clause and insert the following:

SECTION 1. SHORT TITLE

This Act may be cited as the "Drinking Water System Improvement Act of 2017".

SEC. 2. IMPROVED CONSUMER CONFIDENCE REPORTS.

Section 1414(c)(4) of the Safe Drinking Water Act (42 U.S.C. 300g–3(c)(4)) is amended—

* This is an edited reformatted and augmented version of Report of the Committee on Energy and Commerce United States Senate, Report No. 115–380, dated November 1, 2017.

(1) in the heading for subparagraph (A), by striking “Annual Reports” and inserting “Reports”;

(2) in subparagraph (A), by inserting “, or provide by electronic means,” after “to mail”;

(3) in subparagraph (B)—

(A) in clause (iv), by striking “the Administrator, and” and inserting “the Administrator, including corrosion control efforts, and”; and

(B) by adding at the end the following clause:

“(vii) Identification of, if any—

“(I) exceedances described in paragraph (1)(D) for which corrective action has been required by the Administrator or the State (in the case of a State exercising primary enforcement responsibility for public water systems) during the monitoring period covered by the consumer confidence report; and

“(II) violations that occurred during the monitoring period covered by the consumer confidence report.”; and

(4) by adding at the end the following new subparagraph:

“F. Revisions.—

“(i) Understandability and Frequency.—Not later than 24 months after the Drinking Water System Improvement Act of 2017, the Administrator, in consultation with the parties identified in subparagraph (A), shall issue revisions to the regulations issued under subparagraph (A)—

“(I) to increase—

“(aa) the readability, clarity, and understandability of the information presented in consumer confidence reports; and

“(bb) the accuracy of information presented, and risk communication, in consumer confidence reports; and

“(II) with respect to community water systems that serve 10,000 or more persons, to require each such community water system to provide, by mail, electronic means, or other methods described in clause (ii), a consumer confidence report to each customer of the system at least biannually.

“(ii) Electronic Delivery.—Any revision of regulations pursuant to clause (i) shall allow delivery of consumer confidence reports by methods consistent with methods described in the memorandum ‘Safe Drinking Water Act–Consumer Confidence Report Rule Delivery Options’ issued by the Environmental Protection Agency on January 3, 2013.”.

SEC. 3. CONTRACTUAL AGREEMENTS

(a) In General.—Section 1414(h)(1) of the Safe Drinking Water Act (42 U.S.C. 300g–3(h)(1)) is amended—
   (1) in subparagraph (B), by striking "or" after the semicolon;
   (2) in subparagraph (C), by striking the period at the end and inserting "; or"; and
   (3) by adding at the end the following new subparagraph: "(D) entering into a contractual agreement for significant management or administrative functions of the system to correct violations identified in the plan.".

(b) Technical Amendment.—Section 1414(i)(1) of the Safe Drinking Water Act (42 U.S.C. 300g–3(i)(1)) is amended by inserting a comma after "1417".

SEC. 4. CONSOLIDATION

(a) Mandatory Assessment and Consolidation.—Subsection (h) of section 1414 of the Safe Drinking Water Act (42 U.S.C. 300g–3) is amended by adding at the end the following:

"(3) Authority for Mandatory Assessment and Mandatory Consolidation.—

"(A) Mandatory Assessment.—A State with primary enforcement responsibility or the Administrator (if the State does not have primary enforcement responsibility) may require the owner or operator of a public water system to assess options for consolidation, or transfer of ownership of the system, as described in paragraph (1), if—

"(i) the public water system—

"(I) has repeatedly violated one or more national primary drinking water regulations and such repeated violations are likely to adversely affect human health; and

"(II)

(aa) is unable or unwilling to take feasible and affordable actions, as identified by the State with primary enforcement responsibility or the Administrator (if the State does not have primary enforcement responsibility), that will result in the public water system complying with the national primary drinking water regulations described in subclause (I), including accessing technical assistance and financial assistance through the State loan fund pursuant to section 1452; or

"(bb) has already undertaken actions described in item (aa) without achieving compliance;

"(ii) such consolidation or transfer is feasible; and

"(iii) such consolidation or transfer could result in greater compliance with national primary drinking water regulations.

"(B) Mandatory Consolidation.—After review of an assessment under subparagraph (A), a State with primary enforcement responsibility or the Administrator (if the State does not have primary enforcement responsibility) may require the owner or operator of a public water system that completed such assessment to submit a plan for consolidation, or transfer of ownership of the system, under paragraph (1), and complete the actions required under such plan if—

"(i) the owner or operator of the public water system—

"(I) has not taken steps to complete consolidation;

"(II) has not transferred ownership of the system; or

"(III) was unable to achieve compliance after taking the actions described in clause (i)(II)(aa) of subparagraph (A);

"(ii) since completing such assessment, the public water system has violated one or more national primary drinking water regulations and such violations are likely to adversely affect human health; and

"(iii) such consolidation or transfer is feasible.

"(4) Financial Assistance.—Notwithstanding section 1452(a)(3), a public water system undertaking consolidation or transfer of ownership or alternative actions to achieve compliance pursuant to this subsection may receive assistance under section 1452 to carry out such consolidation, transfer, or alternative actions.

"(5)Protection of Nonresponsible System.—

"(A) Identification of Liabilities.—

"(i) In General.—An owner or operator of a public water system submitting a plan pursuant to paragraph (3) shall identify as part of such plan—

"(I) any potential liability for damages arising from each specific violation identified in the plan of which the owner or operator is aware; and

"(II) any funds or other assets that are available to satisfy such liability, as of the date of submission of such plan, to the public water system that committed such violation.

"(ii) Inclusion.—In carrying out clause (i), the owner or operator shall take reasonable steps to ensure that all potential liabilities for damages arising from each specific violation identified in the plan submitted pursuant to paragraph (3) are identified.

"(B) Reservation of Funds.—A public water system that has completed the actions required under a plan submitted and approved pursuant to paragraph (3) shall not be liable under this title for a violation of this title identified in the plan, except to the extent to which funds or other assets are identified pursuant to subparagraph (A)(i)(II) as available to satisfy such liability.

"(6) Regulations.—Not later than 2 years after the date of enactment of the Drinking Water System Improvement Act of 2017, the Administrator shall promulgate regulations to implement paragraphs (3), (4), and (5).".

(b) Retention of Primary Enforcement Authority.—

(1) In General.—Section 1413(a) of the Safe Drinking Water Act (42 U.S.C. 300g–2(a)) is amended—

(A) in paragraph (5), by striking "; and" and inserting a semicolon;

(B) by redesignating paragraph (6) as paragraph (7); and

(C) by inserting after paragraph (5) the following new paragraph:

"(6)has adopted and is implementing procedures for requiring public water systems to assess options for, and complete, consolidation or transfer of ownership, in accordance with the regulations issued by the Administrator under section 1414(h)(6); and".

(2) Conforming Amendment.—Section 1413(b)(1) of the Safe Drinking Water Act (42 U.S.C. 300g–2(b)(1)) is amended by striking "of paragraphs (1), (2), (3), and (4)".

SEC. 5. IMPROVED ACCURACY AND AVAILABILITY OF COMPLIANCE MONITORING DATA.

Section 1414 of the Safe Drinking Water Act (42 U.S.C. 300g–3) is amended by adding at the end the following new subsection:

"(j) Improved Accuracy and Availability of Compliance Monitoring Data.—

"(1)Strategic Plan.—Not later than 1 year after the date of enactment of this subsection, the Administrator, in coordination with States, public water systems, and other interested stakeholders, shall develop and provide to Congress a strategic plan for improving the accuracy and availability of monitoring data collected to demonstrate compliance with national primary drinking water regulations and submitted—

"(A) by public water systems to States; or

"(B) by States to the Administrator.

"(2)Evaluation.—In developing the strategic plan under paragraph (1), the Administrator shall evaluate any challenges faced—

"(A) in ensuring the accuracy and integrity of submitted data described in paragraph (1);

"(B) by States and public water systems in implementing an electronic system for submitting such data, including the technical and economic feasibility of implementing such a system; and

"(C) by users of such electronic systems in being able to access such data.

"(3) Findings and Recommendations.—The Administrator shall include in the strategic plan provided to Congress under paragraph (1)—

"(A) a summary of the findings of the evaluation under paragraph (2); and

"(B) recommendations on practicable, cost-effective methods and means that can be employed to improve the accuracy and availability of submitted data described in paragraph (1).

"(4) Consultation.—In developing the strategic plan under paragraph (1), the Administrator may, as appropriate, consult with States or other Federal agencies that have experience using practicable methods and means to improve the accuracy and availability of submitted data described in such paragraph.".

SEC. 6. ASSET MANAGEMENT

Section 1420 of the Safe Drinking Water Act (42 U.S.C. 300g–9) is amended—

(1) in subsection (c)(2)—

(A) in subparagraph (D), by striking "; and" and inserting a semicolon;

(B) in subparagraph (E), by striking the period at the end and inserting "; and"; and

(C) by adding at the end the following new subparagraph:

"(F) a description of how the State will, as appropriate—

"(i) encourage development by public water systems of asset management plans that include best practices for asset management; and

"(ii) assist, including through the provision of technical assistance, public water systems in training operators or other relevant and appropriate persons in implementing such asset management plans.";

(2) in subsection (c)(3), by inserting ", including efforts of the State to encourage development by public water systems of asset management plans and to assist public water systems in training relevant and appropriate persons in implementing such asset management plans" after "public water systems in the State"; and

(3) in subsection (d), by adding at the end the following new paragraph:

"(5) INFORMATION ON ASSET MANAGEMENT PRACTICES.—Not later than 5 years after the date of enactment of this paragraph, and not less often than every 5 years thereafter, the Administrator shall review and, if appropriate, update educational materials, including handbooks, training materials, and technical information, made available by the Administrator to owners, managers, and operators of public water systems, local officials, technical assistance providers (including nonprofit water associations), and State personnel concerning best practices for asset management strategies that may be used by public water systems.".

SEC. 7. COMMUNITY WATER SYSTEM RISK AND RESILIENCE

(a) In General.—Section 1433 of the Safe Drinking Water Act (42 U.S.C. 300i–2) is amended to read as follows:

Sec. 1433. Community Water System Risk and Resilience

"(a) Risk and Resilience Assessments.—

"(1) In General.—Each community water system serving a population of greater than 3,300 persons shall conduct an assessment of the risks to, and resilience of, its system. Such an assessment—

"(A) shall include an assessment of—

"(i) the risk to the system from malevolent acts and natural hazards;

"(ii) the resilience of the pipes and constructed conveyances, physical barriers, source water, water collection and intake, pretreatment, treatment, storage and distribution facilities, electronic, computer, or other automated systems (including the security of such systems) which are utilized by the system;

"(iii) the monitoring practices of the system;

"(iv) the financial infrastructure of the system;

"(v) the use, storage, or handling of various chemicals by the system; and

"(vi) the operation and maintenance of the system; and

"(B) may include an evaluation of capital and operational needs for risk and resilience management for the system.

"(2) Baseline Information.—The Administrator, not later than August 1, 2019, after consultation with appropriate departments and agencies of the Federal Government and with State and local governments, shall provide baseline information on malevolent acts of relevance to community water systems, which shall include consideration of acts that may—

"(A) substantially disrupt the ability of the system to provide a safe and reliable supply of drinking water; or

"(B) otherwise present significant public health or economic concerns to the community served by the system.

"(3) Certification.—

"(A) Certification.—Each community water system described in paragraph (1) shall submit to the Administrator a certification that the system has conducted an assessment complying with paragraph (1). Such certification shall be made prior to—

"(i) March 31, 2020, in the case of systems serving a population of 100,000 or more;

"(ii) December 31, 2020, in the case of systems serving a population of 50,000 or more but less than 100,000; and

"(iii) June 30, 2021, in the case of systems serving a population greater than 3,300 but less than 50,000.

"(B)Review and Revision.—Each community water system described in paragraph (1) shall review the assessment of such system conducted under such paragraph at least once every 5 years after the applicable deadline for submission of its certification under subparagraph (A) to determine whether such assessment should be revised. Upon completion of such a review, the community water system shall submit to the Administrator a certification that the system has reviewed its assessment and, if applicable, revised such assessment.

"(4)Contents of Certifications.—A certification required under paragraph (3) shall contain only—

"(A)information that identifies the community water system submitting the certification;

"(B)the date of the certification; and

"(C)a statement that the community water system has conducted, reviewed, or revised the assessment, as applicable.

"(5)Provision to Other Entities.—No community water system shall be required under State or local law to provide an assessment described in this section (or revision thereof) to any State, regional, or local governmental entity solely by reason of the requirement set forth in paragraph (3) that the system submit a certification to the Administrator.

"(b)Emergency Response Plan.—Each community water system serving a population greater than 3,300 shall prepare or revise, where necessary, an emergency response plan that incorporates findings of the assessment conducted under subsection (a) for such system (and any revisions thereto). Each community water system shall certify to the Administrator, as soon as reasonably possible after the date of enactment of the Drinking Water System Improvement Act of 2017, but not later than 6 months after completion of the assessment under subsection (a), that the system has completed such plan. The emergency response plan shall include—

"(1)strategies and resources to improve the resilience of the system, including the physical security and cybersecurity of the system;

"(2)plans and procedures that can be implemented, and identification of equipment that can be utilized, in the event of a malevolent act or natural hazard that threatens the ability of the community water system to deliver safe drinking water;

"(3)actions, procedures, and equipment which can obviate or significantly lessen the impact of a malevolent act or natural hazard on the public health and the safety and supply of drinking water provided to communities and individuals, including the development of alternative source water options, relocation of water intakes, and construction of flood protection barriers; and

"(4)strategies that can be used to aid in the detection of malevolent acts or natural hazards that threaten the security or resilience of the system.

"(c)Coordination.—Community water systems shall, to the extent possible, coordinate with existing local emergency planning committees established pursuant to the Emergency Planning and Community Right-To-Know Act of 1986 (42 U.S.C. 11001 et seq.) when preparing or revising an assessment or emergency response plan under this section.

"(d)Record Maintenance.—Each community water system shall maintain a copy of the assessment conducted under subsection (a) and the emergency response plan prepared under subsection (b) (including any revised assessment or plan) for 5 years after the date on which a certification of such assessment or plan is submitted to the Administrator under this section.

"(e)Guidance to Small Public Water Systems.—The Administrator shall provide guidance and technical assistance to community water systems serving a population of less than 3,300 persons on how to conduct resilience assessments, prepare emergency response plans, and address threats from malevolent acts and natural hazards that threaten to disrupt the provision of safe drinking water or significantly affect the public health or significantly affect the safety or supply of drinking water provided to communities and individuals.

"(f) Alternative Preparedness and Operational Resilience Programs.—

"(1)Satisfaction of Requirement.—A community water system that is required to comply with the requirements of subsections (a) and (b) may satisfy such requirements by—

"(A) using and complying with technical standards that the Administrator has recognized under paragraph (2); and

"(B) submitting to the Administrator a certification that the community water system is complying with subparagraph (A).

"(2) Authority to Recognize.—Consistent with section 12(d) of the National Technology Transfer and Advancement Act of 1995, the Administrator shall recognize technical standards that are developed or adopted by third-party organizations or voluntary consensus standards bodies that carry out the objectives or activities required by this section as a means of satisfying the requirements under subsection (a) or (b).

"(g) Technical Assistance and Grants.—

"(1)In General.—The Administrator shall establish and implement a program, to be known as the Drinking Water Infrastructure Risk and Resilience Program, under which the Administrator may award grants in each of fiscal years 2018 through 2022 to owners or operators of community water systems for the purpose of increasing the resilience of such community water systems.

"(2)Use of Funds.—As a condition on receipt of a grant under this section, an owner or operator of a community water system shall agree to use the grant funds exclusively to assist in the planning, design, construction, or implementation of a program or project consistent with an emergency response plan prepared pursuant to subsection (b), which may include—

"(A) the purchase and installation of equipment for detection of drinking water contaminants or malevolent acts;

"(B) the purchase and installation of fencing, gating, lighting, or security cameras;

"(C)the tamper-proofing of manhole covers, fire hydrants, and valve boxes;

"(D)the purchase and installation of improved treatment technologies and equipment to improve the resilience of the system;

"(E)improvements to electronic, computer, financial, or other automated systems and remote systems;

"(F)participation in training programs, and the purchase of training manuals and guidance materials, relating to security and resilience;

"(G)improvements in the use, storage, or handling of chemicals by the community water system;

"(H)security screening of employees or contractor support services;

"(I) equipment necessary to support emergency power or water supply, including standby and mobile sources; and

"(J) the development of alternative source water options, relocation of water intakes, and construction of flood protection barriers.

"(3)Exclusions.—A grant under this subsection may not be used for personnel costs, or for monitoring, operation, or maintenance of facilities, equipment, or systems.

"(4)Technical Assistance.—For each fiscal year, the Administrator may use not more than $5,000,000 from the funds made available to carry out this subsection to provide technical assistance to community water systems to assist in responding to and alleviating a vulnerability that would substantially disrupt the ability of the system to provide a safe and reliable supply of drinking water (including sources of water for such systems) which the Administrator determines to present an immediate and urgent need.

"(5)Grants for Small Systems.—For each fiscal year, the Administrator may use not more than $10,000,000 from the funds made available to carry out this subsection to make grants to community water systems serving a population of less than 3,300 persons, or nonprofit organizations receiving assistance under section 1442(e), for activities and projects undertaken in accordance with the guidance provided to such systems under subsection (e) of this section.

"(6)Authorization of Appropriations.—To carry out this subsection, there are authorized to be appropriated $35,000,000 for each of fiscal years 2018 through 2022.

"(h) Definitions.—In this section—

"(1)the term 'resilience' means the ability of a community water system or an asset of a community water system to adapt to or withstand the effects of a malevolent act or natural hazard without interruption to the asset's or system's function, or if the function is interrupted, to rapidly return to a normal operating condition; and

"(2)the term 'natural hazard' means a natural event that threatens the functioning of a community water system, including an earthquake, tornado, flood, hurricane, wildfire, and hydrologic changes.".

(b) Sensitive Information.—

(1) Protection from Disclosure.—Information submitted to the Administrator of the Environmental Protection Agency pursuant to section 1433 of the Safe Drinking Water Act, as in effect on the day before the date of enactment of the Drinking Water System Improvement Act of 2017, shall be protected from disclosure in accordance with the provisions of such section as in effect on such day.

(2) Disposal.—The Administrator, in partnership with community water systems (as defined in section 1401 of the Safe Drinking Water Act), shall develop a strategy to, in a timeframe determined appropriate by the Administrator, securely and permanently dispose of, or return to the applicable community water system, any information described in paragraph (1).

SEC. 8. AUTHORIZATION FOR GRANTS FOR STATE PROGRAMS

Section 1443(a)(7) of the Safe Drinking Water Act (42 U.S.C. 300j–2(a)(7)) is amended by striking "$100,000,000 for each of fiscal years 1997 through 2003" and inserting "$150,000,000 for each of fiscal years 2018 through 2022".

SEC. 9. MONITORING FOR UNREGULATED CONTAMINANTS.

(a) In General.—Section 1445 of the Safe Drinking Water Act (42 U.S.C. 300j–4) is amended by adding at the end the following:

"(j) Monitoring by Certain Systems.—

"(1)In General.—Notwithstanding subsection (a)(2)(A), the Administrator shall, subject to the availability of appropriations for such purpose—

"(A)require public water systems serving between 3,300 and 10,000 persons to monitor for unregulated contaminants in accordance with this section; and

"(B)ensure that only a representative sample of public water systems serving less than 3,300 persons are required to monitor.

"(2)Effective Date.—Paragraph (1) shall take effect 3 years after the date of enactment of this subsection.

"(3)Limitation.—Paragraph (1) shall take effect unless the Administrator determines that there is not sufficient laboratory capacity to accommodate the analysis necessary to carry out monitoring required under such paragraph.

"(4) Authorization of Appropriations.—There are authorized to be appropriated $15,000,000 in each fiscal year for which monitoring is required to be carried out under this subsection for the Administrator to pay the reasonable cost of such testing and laboratory analysis as are necessary to carry out monitoring required under this subsection.".

(b) Authorization of Appropriations.—Section 1445(a)(2)(H) of the Safe Drinking Water Act (42 U.S.C. 300j–4(a)(2)(H)) is amended by striking "1997 through 2003" and inserting "2018 through 2022".

(c) Inclusion in Data Base.—Section 1445(g)(7) of the Safe Drinking Water Act (42 U.S.C. 300j–4(g)(7)) is amended by—

(1) striking "and" at the end of subparagraph (B);

(2) redesignating subparagraph (C) as subparagraph (D); and

(3) inserting after subparagraph (B) the following:

"(C) if applicable, monitoring information collected by public water systems pursuant to subsection (j) that is not duplicative of monitoring information included in the data base under subparagraph (B) or (D); and".

SEC. 10. STATE REVOLVING LOAN FUNDS

(a) Use of Funds.—Section 1452(a)(2)(B) of the Safe Drinking Water Act (42 U.S.C. 300j–12(a)(2)(B)) is amended by striking "(including expenditures for planning, design, and associated preconstruction activities, including activities relating to the siting of the facility, but not" and inserting "(including expenditures for planning, design, siting, and associated preconstruction activities, or for replacing or rehabilitating aging treatment, storage, or distribution facilities of public water systems, but not".

(b) American Iron and Steel Products.—Section 1452(a)(4)(A) of the Safe Drinking Water Act (42 U.S.C. 300j–12(a)(4)(A)) is amended by striking "fiscal year 2017" and inserting "fiscal years 2018 through 2022".

(c) Evaluation.—Section 1452(a) of the Safe Drinking Water Act (42 U.S.C. 300j–12(a)) is amended by adding at the end the following:

"(5)Evaluation.—During fiscal years 2018 through 2022, a State may provide financial assistance under this section to a public water system serving a population of more than 10,000 for an expenditure described in paragraph (2) only if the public water system—

"(A) considers the cost and effectiveness of relevant processes, materials, techniques, and technologies for carrying out the project or activity that is the subject of the expenditure; and

"(B) certifies to the State, in a form and manner determined by the State, that the public water system has made such consideration.".

(d) Prevailing Wages.—Section 1452(a) of the Safe Drinking Water Act (42 U.S.C. 300j–12(a)) is further amended by adding at the end the following:

"(6)Prevailing Wages.—The requirements of section 1450(e) shall apply to any construction project carried out in whole or in part with assistance made available by a drinking water treatment revolving loan fund.".

(e) Assistance for Disadvantaged Communities.—Section 1452(d)(2) of the Safe Drinking Water Act (42 U.S.C. 300j–12(d)(2)) is amended to read as follows:

"(2)Total Amount Of Subsidies.—For each fiscal year, of the amount of the capitalization grant received by the State for the year, the total amount of loan subsidies made by a State pursuant to paragraph (1)—

"(A) may not exceed 35 percent; and

"(B) to the extent that there are sufficient applications for loans to communities described in paragraph (1), may not be less than 6 percent.".

(f) Types of Assistance.—Section 1452(f)(1) of the Safe Drinking Water Act (42 U.S.C. 300j–12(f)(1)) is amended—

(1) by redesignating subparagraphs (C) and (D) as subparagraphs (D) and (E), respectively;

(2) by inserting after subparagraph (B) the following new subparagraph:

"(C) each loan will be fully amortized not later than 30 years after the completion of the project, except that in the case of a disadvantaged community (as defined in subsection (d)(3)) a State may provide an extended term for a loan, if the extended term—

"(i) terminates not later than the date that is 40 years after the date of project completion; and

"(ii) does not exceed the expected design life of the project;"; and

(3) in subparagraph (B), by striking "1 year after completion of the project for which the loan was made" and all that follows through "design life of the project;" and inserting "18 months after completion of the project for which the loan was made;".

(g) Needs Survey.—Section 1452(h) of the Safe Drinking Water Act (42 U.S.C. 300j–12(h)) is amended—

(1) by striking "The Administrator" and inserting "(1) The Administrator"; and

(2) by adding at the end the following new paragraph:

"(2) Any assessment conducted under paragraph (1) after the date of enactment of the Drinking Water System Improvement Act of 2017 shall include an assessment of costs to replace all lead service lines (as defined in section 1459B(a)(4)) of all eligible public water systems in the United States, and such assessment shall describe separately the costs associated with replacing the portions of such lead service lines that are owned by an eligible public water system and the costs associated with replacing any remaining portions of such lead service lines, to the extent practicable.".

(h) Other Authorized Activities.—Section 1452(k)(1)(C) of the Safe Drinking Water Act (42 U.S.C. 300j–12(k)(1)(C)) is amended by striking "for fiscal years 1996 and 1997 to delineate and assess source water protection areas in accordance with section 1453" and inserting "to delineate, assess, and update assessments for source water protection areas in accordance with section 1453".

(i) Authorization for Capitalization Grants to States for State Drinking Water Treatment Revolving Loan Funds.—Section 1452(m) of the Safe Drinking Water Act (42 U.S.C. 300j–12(m)) is amended—

(1) by striking the first sentence and inserting the following:

"(1)There are authorized to be appropriated to carry out the purposes of this section—

"(A) $1,200,000,000 for fiscal year 2018;

"(B) $1,400,000,000 for fiscal year 2019;

"(C) $1,600,000,000 for fiscal year 2020;

"(D) $1,800,000,000 for fiscal year 2021; and

"(E) $2,000,000,000 for fiscal year 2022.";

(2) by striking "To the extent amounts authorized to be" and inserting the following:

"(2) To the extent amounts authorized to be"; and

(3) by striking "(prior to the fiscal year 2004)".

(j) Best Practices for Administration Of State Revolving Loan Funds.—Section 1452 of the Safe Drinking Water Act (42 U.S.C. 300j–12) is amended by adding after subsection (r) the following:

"(s) Best Practices for State Loan Fund Administration.—The Administrator shall—

"(1)collect information from States on administration of State loan funds established pursuant to subsection (a)(1), including—

"(A) efforts to streamline the process for applying for assistance through such State loan funds;

"(B) programs in place to assist with the completion of applications for assistance through such State loan funds;

"(C) incentives provided to public water systems that partner with small public water systems to assist with the application process for assistance through such State loan funds;

"(D) practices to ensure that amounts in such State loan funds are used to provide loans, loan guarantees, or other authorized assistance in a timely fashion;

"(E) practices that support effective management of such State loan funds;

"(F) practices and tools to enhance financial management of such State loan funds; and

"(G) key financial measures for use in evaluating State loan fund operations, including—

"(i) measures of lending capacity, such as current assets and current liabilities or undisbursed loan assistance liability; and

"(ii) measures of growth or sustainability, such as return on net interest;

"(2) not later than 3 years after the date of enactment of the Drinking Water System Improvement Act of 2017, disseminate to the States best practices for administration of such State loan funds, based on the information collected pursuant to this subsection; and

"(3) periodically update such best practices, as appropriate.".

SEC. 11. AUTHORIZATION FOR SOURCE WATER PETITION PROGRAMS

Section 1454(e) of the Safe Drinking Water Act (42 U.S.C. 300j–14(e)) is amended by striking "1997 through 2003" and inserting "2018 through 2022".

SEC. 12. REVIEW OF TECHNOLOGIES

Part E of the Safe Drinking Water Act (42 U.S.C. 300j et seq.) is amended by adding at the end the following new section:

Sec. 1459C. Review of Technologies

"(a) Review.—The Administrator, after consultation with appropriate departments and agencies of the Federal Government and with State and local governments, shall review (or enter into contracts or cooperative agreements to provide for a review of) existing and potential methods, means, equipment, and technologies (including review of cost, availability, and efficacy of such methods, means, equipment, and technologies) that—

"(1) ensure the physical integrity of community water systems;

"(2) prevent, detect, and respond to any contaminant for which a national primary drinking water regulation has been promulgated in community water systems and source water for community water systems;

"(3) allow for use of alternate drinking water supplies from nontraditional sources; and

"(4) facilitate source water assessment and protection.

"(b) Inclusions.—The review under subsection (a) shall include review of methods, means, equipment, and technologies—

"(1) that are used for corrosion protection, metering, leak detection, or protection against water loss;

"(2) that are intelligent systems, including hardware, software, or other technology, used to assist in protection and detection described in paragraph (1);

"(3) that are point-of-use devices or point-of-entry devices;

"(4) that are physical or electronic systems that monitor, or assist in monitoring, contaminants in drinking water in real-time; and

"(5) that allow for the use of nontraditional sources for drinking water, including physical separation and chemical and biological transformation technologies.

"(c) Availability.—The Administrator shall make the results of the review under subsection (a) available to the public.

"(d) Authorization of Appropriations.—There are authorized to be appropriated to the Administrator to carry out this section $10,000,000 for fiscal year 2018, which shall remain available until expended.".

SEC. 13. DRINKING WATER FOUNTAIN REPLACEMENT FOR SCHOOLS

(a) In General.—Part F of the Safe Drinking Water Act (42 U.S.C. 300j–21 et seq.) is amended by adding at the end the following:

Sec. 1465. Drinking Water Fountain Replacement for Schools

"(a) Establishment.—Not later than 1 year after the date of enactment of this section, the Administrator shall establish a grant program to provide assistance to local educational agencies for the replacement of drinking water fountains manufactured prior to 1988.

"(b) Use of Funds.—Funds awarded under the grant program—

"(1) shall be used to pay the costs of replacement of drinking water fountains in schools; and

"(2) may be used to pay the costs of monitoring and reporting of lead levels in the drinking water of schools of a local educational agency receiving such funds, as determined appropriate by the Administrator.

"(c) Priority.—In awarding funds under the grant program, the Administrator shall give priority to local educational agencies based on economic need.

"(d) Authorization of Appropriations.—There are authorized to be appropriated to carry out this section not more than $5,000,000 for each of fiscal years 2018 through 2022.".

(b) Definitions.—Section 1461(5) of the Safe Drinking Water Act (42 U.S.C. 300j–21(5)) is amended by inserting "or drinking water fountain" after "water cooler" each place it appears.

SEC. 14. SOURCE WATER

(a) Addressing Source Water Used for Drinking Water.—Section 304 of the Emergency Planning and Community Right-To-Know Act of 1986 (42 U.S.C. 11004) is amended—

(1) in subsection (b)(1), by striking "State emergency planning commission" and inserting "State emergency response commission"; and

(2) by adding at the end the following new subsection:

"(e) Addressing Source Water Used for Drinking Water.—

"(1) Applicable State Agency Notification.—A State emergency response commission shall—

"(A) promptly notify the applicable State agency of any release that requires notice under subsection (a);

"(B) provide to the applicable State agency the information identified in subsection (b)(2); and

"(C) provide to the applicable State agency a written followup emergency notice in accordance with subsection (c).

"(2)Community Water System Notification.—

"(A) In General.—An applicable State agency receiving notice of a release under paragraph (1) shall—

"(i) promptly forward such notice to any community water system the source waters of which are affected by the release;

"(ii) forward to the community water system the information provided under paragraph (1)(B); and

"(iii) forward to the community water system the written followup emergency notice provided under paragraph (1)(C).

"(B) Direct Notification.—In the case of a State that does not have an applicable State agency, the State emergency response commission shall provide the notices and information described in paragraph (1) directly to any community water system the source waters of which are affected by a release that requires notice under subsection (a).

"(3)Definitions.—In this subsection:

"(A) Community Water System.—The term 'community water system' has the meaning given such term in section 1401(15) of the Safe Drinking Water Act.

"(B) Applicable State Agency.—The term 'applicable State agency' means the State agency that has primary responsibility to enforce the requirements of the Safe Drinking Water Act in the State.".

(b) Availability to Community Water Systems.—Section 312(e) of the Emergency Planning and Community Right-To-Know Act of 1986 (42 U.S.C. 11022(e)) is amended—

(1) in paragraph (1), by striking "State emergency planning commission" and inserting "State emergency response commission"; and

(2) by adding at the end the following new paragraph:

"(4) Availability to Community Water Systems.—

"(A) In General.—An affected community water system may have access to tier II information by submitting a request to the State emergency response commission or the local emergency planning committee. Upon receipt of a request for tier II information, the State commission or local committee shall, pursuant to paragraph (1), request the facility owner or operator for the tier II information and make available such information to the affected community water system.

"(B) Definition.—In this paragraph, the term 'affected community water system' means a community water system (as defined in section 1401(15) of the Safe Drinking Water Act) that receives supplies of drinking water from a source water area, delineated under section 1453 of the Safe Drinking Water Act, in which a facility that is required to prepare and submit an inventory form under subsection (a)(1) is located.".

SEC. 15. REPORT ON FEDERAL CROSS-CUTTING REQUIREMENTS

(a) Report.—Not later than one year after the date of enactment of this Act, the Comptroller General shall submit to Congress a report containing the results of a study, to be conducted in consultation with the Administrator of the Environmental Protection Agency, any State agency that has primary responsibility to enforce the requirements of the Safe Drinking Water Act (42 U.S.C. 300f et seq.) in a State, and public water systems, to identify demonstrations of compliance with a State or local environmental law that may be substantially equivalent to any demonstration required by the Administrator for compliance with a Federal cross-cutting requirement.

(b) Definitions.—In this subsection:

(1) Federal Cross-Cutting Requirement.—The term "Federal cross-cutting requirement" means a requirement of a Federal law or regulation, compliance with which is a condition on receipt of a loan or loan guarantee pursuant to section 1452 of the Safe Drinking Water Act (42 U.S.C. 300j–12), that, if applied with respect to projects and activities for which a public water system receives such a loan or loan guarantee, would be substantially equivalent to a requirement of an applicable State or local law.

(2) Public Water System.—The term "public water system" has the meaning given that term in section 1401 of the Safe Drinking Water Act (42 U.S.C. 300f).

PURPOSE AND SUMMARY

The purpose of the bill is to amend the Safe Drinking Water Act to improve public water systems and enhance compliance with such Act.

BACKGROUND AND NEED FOR LEGISLATION

The United States uses 42 billion gallons of water a day—treated to meet Federal drinking water standards—to support a variety of needs.[9] According to the Congressional Research Service (CRS), more than 299 million Americans are served by more than 51,300 community water systems (CWSs). Most community water systems (82 percent of all CWSs) are relatively small, serving 3,300 people or fewer; but these systems provide water to just 9 percent of the total population served by community water systems. In contrast, 8 percent of all CWSs serve 82 percent of the population served.[10]

Treated drinking water is delivered across the United States, via one million miles of pipes, by privately and publicly owned water systems. Many of these pipes were laid in the early to mid-20th century and have a lifespan of 75 to 100 years.[11] While the American Society of Civil Engineers (ASCE) reports the quality of drinking water in the United States remains high, ASCE and others also spotlight concerns directly related to water system integrity, efficiency, and affordability. Specifically, they point to an estimated 240,000 water main breaks per year in the United States that waste over two trillion gallons of treated drinking water. These leaks waste 14 to 18 percent of treated water per day—an amount that could support 15 million households.[12]

In April 2013, the Environmental Protection Agency (EPA) published its most recent survey of capital improvement needs for drinking water infrastructure. That survey indicated that water systems need to invest $384.2 billion on infrastructure improvements over 20 years (from 2011 to 2030) to ensure the provision of safe tap water.[13] EPA also reported that $42.0 billion (10.9 percent) of reported drinking water system needs are attributable to Safe Drinking Water Act (SDWA) compliance. The remaining 89.1 percent of EPA-identified needs are for projects that are not regulatory, but are needed to meet the Act's health protection objectives. A study by the American Water Works Association (AWWA) projects that restoring infrastructure and expanding water systems to keep up with population growth would require a nationwide investment of at least $1 trillion through 2035.[14]

The Congressional Budget Office reports that, in 2014, the Federal share of total public spending on water and wastewater utilities was 4 percent, while State and local government expenditures accounted for 94 percent of all public spending on this infrastructure.[15]

---

[9] http://www.infrastructurereportcard.org/wp-content/uploads/2017/01/Drinking-Water- Final.pdf

[10] http:// www.crs.gov/ Reports/ RL31243?source=search&guid= c987b8c3502d477b8842999a6ea62e7a&index=10

[11] American Water Works Association, Buried No Longer: Confronting American's Water Infra- structure Challenge, 2012, http://www.awwa.org/legislation-regulation/issues/ infrastructure-financing.aspx.

[12] Op. Cit.

[13] https://www.epa.gov/sites/production/files/2015-07/documents/epa816r13006.pdf

[14] http://www.crs.gov/Reports/RS22037?source=search&guid=923f50f6c1274772996da7f8f1a 3551b&index=6# _Toc466362844

[15] Congressional Budget Office, Public Spending on Transportation and Water Infrastructure, 1956 to 2014, March 2015, p. 28.

User fees, primarily in the form of water utility rates, typically generate funds for daily operation and maintenance and long-term capital investments for drinking water and wastewater systems. Both the EPA and the United Nation's Development Program recommend affordability thresholds for water and wastewater services of 2.5 percent and 3 percent, respectively, of median household income.[16] The average price of treating and distributing water in the United States is about $1.50 for 1,000 gallons—at that price, a gallon of water costs less than one penny.[17] While the AWWA estimates that drinking water rates, annualized from 2004 to 2014, have increased 5.5 percent, AWWA also shows that water rates have dropped three percent between 2012 and 2014.[18]

However, an ongoing problem for local water systems is how to finance major projects—increasing rates, borrowing on the private market, seeking Federal or State assistance, or some combination of these.

Safe Drinking Water Act

The Safe Drinking Water Act (SDWA) not only contains Federal authority for regulating contaminants in drinking water delivery systems, it also includes the Drinking Water State Revolving Fund (DWSRF) program.[19] The DWSRF was created by Congress in the 1996 SDWA Amendments to provide financing for infrastructure improvements at drinking water systems. Congress envisioned a program operating in perpetuity from which the principal and interest payments on old loans would be used to issue new loans, and from which a portion of each State's allotment could be set aside for State drinking water agencies to provide regulatory oversight and direct assistance to water systems.[20]

Specifically, the DWSRF program permits EPA to make grants to States to capitalize DWSRFs, which States may then use to make low-interest loans to public water systems (PWSs) for activities EPA determines facilitate compliance or significantly further the SDWA's health protection objectives. States must match 20 percent of the Federal grant. Grants are allotted based on the results of needs surveys issued quadrennially by EPA. Each State and the District of Columbia must receive at least 1 percent of the appropriated funds.[21]

[16] http://www2.pacinst.org/wp-content/uploads/2013/01/water-rates-affordability.pdf

[17] https://www.fcwa.org/story_of_water/html/costs.htm

[18] https://www.awwa.org/resources-tools/water-and-wastewater-utility-management/water- wastewater-rates.aspx

[19] SDWA § 1412 and § 1452

[20] http://www.asdwa.org/document/docWindow.cfm?fuseaction=document.viewDocument& documentid=2683& documentFormatId=3404

[21] http://www.crs.gov/Reports/RL31243?source=search&guid=c987b8c3502d477b8842999 a6ea62e7a&index=10 #_Toc476131535.

In addition, States must make available 15 percent of their annual DWSRF allotment for loan assistance to systems that serve 10,000 or fewer persons to the extent that there are systems of that size within a State applying for funding of qualifying activities. States may also use up to 30 percent of their DWSRF grant to provide loan subsidies (including forgiveness of principal) to help economically disadvantaged communities. Finally, States may also use up to 4 percent of funds for technical assistance, source water protection and capacity development programs, and operator certification.[22]

When last reauthorized in 1996, SDWA authorized appropriations of $599 million for fiscal year 1994 and $1 billion per year for fiscal year 1995 through fiscal year 2003 for DWSRF capitalization grants. Of those amounts, EPA was either directed or given the ability to reserve, from annual DWSRF appropriations, 0.33 percent for financial assistance to territories and 1.5 percent for Indian tribes and Alaska Native Villages, $10 million for health effects research on drinking water contaminants, $2 million for the costs of monitoring for unregulated contaminants, and up to 2 percent for technical assistance. Between fiscal year 1997 and fiscal year 2016, Congress appropriated over $20 billion, and more than 12,400 projects received assistance through the program.[23]

The Water Infrastructure Improvements for the Nation Act (WIIN Act, section 322 of P.L. 114–322) made several amendments to the DWSRF provisions. Among other changes, the amendments increased the portion of the annual DWSRF capitalization grants that States may use to cover program administration costs and authorized $300 million over five years for lead pipe replacement and $300 million over five years for aid to disadvantaged and underserved communities.[24] Further, the WIIN Act amended SDWA to require, with some exceptions, that funds made available from a State DWSRF during fiscal year 2017 may not be used for water system projects unless all iron and steel products to be used in the project are produced in the United States.

Water Infrastructure Finance and Innovation Act

According to CRS, a chronic concern is the need for communities to address drinking water infrastructure requirements that are outside the scope of the DWSRF program since they are unrelated to SDWA compliance.[25] These categories include future growth, on-going rehabilitation, and operation and maintenance of systems. EPA has reported that outdated and deteriorated drinking water infrastructure poses a fundamental long-term threat to drinking water safety and that, in many communities, basic infrastructure costs can far exceed SDWA compliance costs. As reported in EPA's most recent drinking water

[22] SDWA 1452(g).
[23] *Id.*
[24] WIIN 2102–2105.
[25] Op. Cit.

needs assessment, less than 11 percent of the 20-year estimated need is directly related to compliance with SDWA regulations.[26]

Congress enacted the Water Infrastructure Finance and Innovation Act (WIFIA) in June 2014,[27] which authorized a pilot loan guarantee program to test the ability of innovative financing tools to promote increased development of, and private investment in, water infrastructure projects while reducing costs to the Federal government. The pilot program is intended to complement, and not replace, the clean water and drinking water SRF programs. The Act authorized $20 million each for fiscal year 2015 and $25 million each for fiscal year 2016 to the Secretary of the Interior and the EPA Administrator, with amounts increasing annually to $50 million each for fiscal year 2019.

Eligible projects include clean water and drinking water SRF-eligible projects and a wide range of water resource development projects that must generally have costs of at least $20.0 million. Such large projects face difficulty securing significant funding through the SRF programs. Moreover, unlike the SRF programs, WIFIA is not focused on regulatory compliance and, therefore, may be more available for other large-scale water infrastructure projects. For projects serving areas with a population of 25,000 or fewer individuals, eligible projects must have a total cost of at least $5 million.

Congress appropriated $20 million in funds for the program in fiscal year 2017. It is estimated that using WIFIA's full financial leveraging ability that a single dollar injected into the program can create $50 dollars for project lending.[28] Under current appropriations, EPA estimates that current budget authority may provide more than $1 billion in credit assistance and may finance over $2 billion in water infrastructure investment.[29]

Clean Water Act SRF

Congress provided States flexibility in setting priorities between the DWSRF and the Clean Water Act SRF (CWSRF) programs to accommodate the divergent drinking water and wastewater needs and priorities among the States. Section 302(a) of the 1996 SDWA Amendments authorized States to transfer as much as 33 percent of the annual DWSRF allotment to the CWSRF or an equivalent amount from the CWSRF to the DWSRF. The Act authorized these transfers through fiscal year 2001. In 2000, EPA recommended that Congress continue to authorize transfers between the SRF programs to give States flexibility to address their most pressing water infrastructure needs.[30]

---

[26] Op. Cit.

[27] (P.L. 113–121, H.R. 3080) includes in Title V, Subtitle C.

[28] http://www.infrastructurereportcard.org/wp-content/uploads/2017/01/Drinking-Water-Final. pdf

[29] *Id.*

[30] http:// www.crs.gov/ Reports/ RS22037? source=search&guid= 923f50f6c 1274772996da7f8f1a3551b& index= 6#ifn2.

Several annual appropriations acts authorized States to continue to transfer as much as 33 percent of funds between the two programs, and in the Department of Interior and Related Agencies Appropriations Act, FY 2006, Congress made this authority permanent.[31]

COMMITTEE ACTION

On May 19, 2017, the Subcommittee on the Environment held a hearing on a Discussion Draft, entitled "Drinking Water System Improvement Act." The Subcommittee received testimony from:

- Lisa Daniels, Director, Bureau of Safe Drinking Water, Pennsylvania Department of Environmental Protection, on behalf of the Association of State Drinking Water Administrators;
- Steve Fletcher, Manager, Washington County Water Company, Nashville, IL, on behalf of the National Rural Water Association;
- Martin Kropelnicki, President and CEO, California Water Service Group, on behalf of the National Association of Water Companies;
- Scott Potter, Director, Nashville Metro Water Services, Nashville, TN, on behalf of the Association of Metropolitan Water Agencies;
- James Proctor, Senior Vice President and General Counsel, McWane, Inc.;
- Lynn Thorp, National Campaigns Director, Clean Water Action; and
- Kurt Vause, Special Projects Director, Anchorage Water and Wastewater Utility, on behalf of the American Water Works Association.

On March 16, 2017, the Subcommittee on the Environment held a hearing entitled, "Reinvestment and Rehabilitation of Our Nation's Drinking Water Delivery Systems." The Subcommittee received testimony from:

- Rudolph Chow, P.E., Director, Baltimore City Department of Public Works, on behalf of the American Municipal Water Association;
- Greg DiLoreto, Chairman, Committee for America's Infrastructure, American Society of Civil Engineers;
- John Donahue, CEO, North Park Public Water District (Machesney Park, IL), on behalf of the American Water Works Association;

[31] The Department of the Interior, Environment, and Related Agencies Appropriations Act, 2006, P.L. 109–54, Title II, August 2, 2005, 119 Stat. 530, provided: "That for fiscal year 2006 and thereafter, State authority under section 302(a) of P.L. 104–182 shall remain in effect."

- Randy Ellingboe, Minnesota Department of Health, on behalf of the Association of State Drinking Water Administrators;
- Martin A. Kropelnicki, President and CEO, California Water Service Group, on behalf of the National Association of Water Companies; and
- Erik Olson, Director, Health and Environment Program, Natural Resources Defense Council.

On July 13, 2017, the Subcommittee on the Environment met in open markup session and forwarded the Discussion Draft, entitled "Drinking Water System Improvement Act," as amended, to the full Committee by a voice vote. On July 25, 2017, the Drinking Water System Improvement Act of 2017 was introduced in the House as H.R. 3387. On July 27, 2017, the full Committee on Energy and Commerce met in open markup session and ordered H.R. 3387, as amended, favorably reported to the House by a voice vote.

Committee Votes

Clause 3(b) of rule XIII requires the Committee to list the record votes on the motion to report legislation and amendments thereto. There were no record votes taken in connection with ordering H.R. 3387 reported.

Oversight Findings and Recommendations

Pursuant to clause 2(b)(1) of rule X and clause 3(c)(1) of rule XIII, the Committee held hearings and made findings that are reflected in this report.

New Budget Authority, Entitlement Authority, and Tax Expenditures

Pursuant to clause 3(c)(2) of rule XIII, the Committee finds that H.R. 3387 would result in no new or increased budget authority, entitlement authority, or tax expenditures or revenues.

CONGRESSIONAL BUDGET OFFICE ESTIMATE

Pursuant to clause 3(c)(3) of rule XIII, the following is the cost estimate provided by the Congressional Budget Office pursuant to section 402 of the Congressional Budget Act of 1974:

U.S. Congress,
Congressional budget office,
Washington, DC, September 22, 2017.

Hon. Greg Walden,
Chairman, Committee on Energy and Commerce,
House of Representatives, Washington, DC.

Dear Mr. Chairman: The Congressional Budget Office has prepared the enclosed cost estimate for H.R. 3387, the Drinking Water System Improvement Act of 2017.

If you wish further details on this estimate, we will be pleased to provide them. The CBO staff contact is Jon Sperl.

Sincerely,
Keith Hall,
Director.

H.R. 3387—Drinking Water System Improvement Act of 2017

Summary: H.R. 3387 would authorize the appropriation of about $9 billion for the Environmental Protection Agency (EPA) to provide grants to public water systems, as well as to state, local, and tribal governments, to support drinking water infrastructure projects and to promote compliance with regulations that implement the Safe Drinking Water Act (SDWA).

CBO estimates that implementing this legislation would cost about $6 billion over the next five years and an additional $3 billion after 2022, assuming appropriation of the authorized amounts.

The staff of the Joint Committee on Taxation (JCT) estimates that enacting the bill would reduce revenues by $572 million over the next 10 years. Because enacting the bill would reduce revenues, pay-as-you-go procedures apply. Enacting the bill would not affect direct spending.

CBO estimates that enacting H.R. 3387 would not increase net direct spending or on-budget deficits by more than $5 billion in any of the four consecutive 10-year periods beginning in 2028.

Table 1. Estimated Budgetary Effects of H.R. 3387

By fiscal year, in millions of dollars—	2017	2018	2019	2020	2021	2022	2023	2024	2025	2026	2027	2017-2022	2017-2027
INCREASES IN SPENDING SUBJECT TO APPROPRIATION													
Estimated Authorization Level	0	1,420	1,610	1,824	2,023	2,222	17	17	17	17	17	9,100	9,186
Estimated Outlays	0	263	805	1,408	1,735	1,923	1,760	1,022	219	24	24	6,135	9,185
DECREASES IN REVENUES													
Estimated Revenues	0	*	– 3	– 10	– 25	– 46	– 72	– 95	– 106	– 108	– 107	– 84	– 572

Sources: CBO and the staff of the Joint Committee on Taxation.

Note: * = between zero and ¥$500,000. Details may not sum to totals because of rounding.

Table 2. Amounts Authorized to be Appropriated for EPA Programs Under H.R. 3387

By fiscal year, in millions of dollars—	2017	2018	2019	2020	2021	2022	2023	2024	2025	2026	2027	2017-2022	2017-2027
Drinking Water SRF Grants:													
Authorization Level	0	1,200	1,400	1,600	1,800	1,000	0	0	0	0	0	8,000	8,000
Estimated Outlays	0	120	620	1,200	1,500	1,700	1,680	980	200	0	0	5,140	8,000
Public Water System Supervision Program													
Authorization Level	0	150	150	150	150	150	0	0	0	0	0	750	750
Estimated Outlays	0	135	150	150	150	150	15	0	0	0	0	735	750
Drinking Water Risk and Resilience Grant Program:													
Authorization Level	0	35	35	35	35	35	0	0	0	0	0	175	175
Estimated Outlays	0	0	11	25	35	35	35	25	11	0	0	105	175

Table 2. (Continued)

By fiscal year, in millions of dollars—												
	2017	2018	2019	2020	2021	2022	2023	2024	2025	2026	2027	2017-2022	2017-2027
Source Water Petition Programs:													
Authorization Level	0	5	5	5	5	5	0	0	0	0	0	25	25
Estimated Outlays	0	0	2	4	5	5	5	4	2	0	0	15	25
Drinking Water Fountain Replacement Grants:													
Authorization Level	0	5	5	5	5	5	0	0	0	0	0	25	25
Estimated Outlays	0	0	2	4	5	5	5	4	2	0	0	15	25
Reauthorize Monitoring for Unregulated Contaminants:													
Authorization Level	0	10	10	10	10	10	0	0	0	0	0	50	50
Estimated Outlays	0	2	15	15	13	5	0	0	0	0	0	50	50
Expand Monitoring for Unregulated Contaminants:													
Authorization Level	0	0	0	15	15	15	15	15	15	15	15	45	120
Estimated Outlays	0	0	0	4	21	21	18	8	4	21	21	45	120
Technology Review:													
Authorization Level	0	10	0	0	0	0	0	0	0	0	0	10	10
Estimated Outlays	0	1	1	4	4	0	0	0	0	0	0	10	10
Other Activities:													
Estimated Authorization Level	0	5	5	4	3	2	2	2	2	2	2	20	31
Estimated Outlays	0	5	5	4	3	2	2	2	2	2	2	20	31
Total Changes:.													
Estimated Authorization Level	0	1,420	1,610	1,824	2,023	2,222	17	17	17	17	17	9,100	9,186
Estimated Outlays	0	263	805	1,408	1,735	1,923	1,760	1,022	219	24	24	6,135	9,185

Note: Details may not sum to totals because of rounding. SRF = State Revolving Fund.

H.R. 3387 would impose intergovernmental and private-sector mandates as defined in the Unfunded Mandates Reform Act (UMRA) on public and private owners and operators of public water systems that are regulated by the SDWA, and on other state and local government entities. Based on information provided by the EPA, public water systems, and state and local agencies, CBO estimates that the total cost of complying with the mandates would fall below the annual thresholds for intergovernmental and private-sector mandates established in UMRA ($78 million and $156 million in 2017, respectively, adjusted annually for inflation).

Estimated cost to the Federal Government: The estimated budgetary effects of the bill are summarized in Table 1. The costs of this legislation fall within budget function 300 (natural resources and environment).

Basis of estimate: For this estimate, CBO assumes that the bill will be enacted near the beginning of fiscal year 2018, that the full amounts authorized or estimated to be necessary will be appropriated for each year, and that outlays will follow historical patterns of spending for existing and similar programs.

Spending Subject to Appropriation

H.R. 3387 would authorize appropriations totaling about $9.1 billion over the 2018–2022 period for the EPA to administer different grant programs that support drinking water infrastructure and help public water systems comply with regulations under the Safe Drinking Water Act (see Table 2).

The bill would authorize the appropriation of $8 billion over the next five years for the EPA to provide capitalization grants for the Drinking Water State Revolving Fund (DWSRF) programs. States use such grants, along with their own funds, to make low-interest loans to communities to build or improve drinking water facilities and infrastructure, and for other projects that improve the quality of drinking water. In addition to reauthorizing federal funding for states' DWSRF programs, the bill also would make several revisions to those programs, including allowing states to direct a greater percentage of funds to disadvantaged communities, extending the repayment terms for loans made by states, and requiring recipients of loans to certify that proposed projects meet certain cost-effectiveness criteria.

H.R. 3387 also would authorize the appropriation of about $1 billion over the next five years for the EPA implement several other grant programs. Specifically, the bill would authorize the appropriation of:

- $750 million for state and tribal agencies to implement programs that enforce compliance with drinking water regulations under the SDWA and provide technical assistance to public water systems;
- $175 million for grants to public water systems to implement projects that mitigate risks to drinking water from natural hazards and security threats;

- $25 million to states to implement partnership programs with public water systems that petition the states for assistance in complying with drinking water regulations;
- $25 million for local educational agencies to pay the costs of replacing drinking water fountains in schools and monitoring for lead contamination;
- $50 million for the EPA to continue funding the laboratory analysis costs of monitoring for unregulated contaminants in drinking water systems;
- Daniels, Dirthe number of small systems that monitor for unregulated contaminants; and
- $10 million in 2018 for the EPA to conduct a comprehensive review of technologies, equipment, and methods that effectively detect and prevent contamination of public drinking water systems.

CBO estimates that the cost to implement the remaining requirements in the bill, (for which the legislation does not specify an authorization level,) would total about $20 million over the next five years, assuming appropriation of the necessary amounts. That funding would be used for various purposes, including providing technical assistance to state agencies, developing guidance and updating tools for risk assessments, conducting a national inventory of any pipes of fittings that are used to connect buildings with the drinking water main supply pipes and are not lead free, and reporting on how water systems can more easily comply with cross-cutting federal, state, and local requirements.

Revenues

H.R. 3387 would authorize the appropriation of $8 billion over the 2018–2022 period for the EPA to make grants to capitalize state revolving loan funds, from which states make loans to finance drinking water infrastructure projects. JCT expects that states would use a portion of those grants to leverage additional funds by issuing tax-exempt bonds. JCT estimates that issuing additional tax-exempt bonds would reduce federal revenues by $572 million over the next 10 years.

Pay-As-You-Go Considerations: The Statutory Pay-As-You-Go Act of 2010 establishes budget-reporting and enforcement procedures for legislation affecting direct spending or revenues. The net changes in revenues that are subject to those pay-as-you-go procedures are shown in the following table.

Increase in long-term direct spending and deficits: CBO estimates that enacting the legislation would not increase net direct spending or on-budget deficits by more than $5 billion in any of the four consecutive 10-year periods beginning in 2028.

CBO Estimate of Pay-As-You-Go Effects For H.R. 3387, As Ordered Reported By the House Committee on Energy and Commerce on July 27, 2017

By fiscal year, in millions of dollars—													
	2017	2018	2019	2020	2021	2022	2023	2024	2025	2026	2027	2017-2022	2017-2027
NET INCREASE IN THE DEFICIT													
Statutory Pay-As-You-Go Impact	0	0	3	10	25	46	72	95	106	108	107	84	572

Intergovernmental and private-sector impact: H.R. 3387 would impose intergovernmental and private-sector mandates as defined in UMRA on public and private owners and operators of public water systems that are regulated by the Safe Drinking Water Act. The bill also would impose intergovernmental mandates on state emergency response commissions (SERCs), local emergency planning committees, and state water agencies that are responsible for notifying the public in the event of a release of hazardous chemicals that affects drinking water. Based on information provided by the EPA, public water systems, and state and local agencies, CBO estimates that the total costs of complying with the mandates would range from $14 million to $36 million per year over the 2018–2022 period for water systems owned by public entities; for water systems owned by private entities, CBO estimates the total costs to comply with the mandates would range from $3 million to $6 million per year over that period. Therefore, CBO estimates that the costs of the mandates would fall below the annual thresholds for intergovernmental and private-sector mandates established in UMRA ($78 million and $156 million in 2017, respectively, adjusted annually for inflation).

Mandates that Apply to Both Public and Private Entities

The bill would impose several mandates on owners and operators of public water systems that are regulated by the Safe Drinking Water Act. Public water systems may be publicly or privately owned. Systems owned by local governments serve the majority of the U.S. population, while many smaller systems are owned by private entities. The bill would require public water systems that serve populations larger than 10,000 to send consumer confidence reports to their customers twice per year; under current law, those systems must send reports once per year. In addition, the bill would require all public water systems to include information in consumer confidence reports about actions taken to control corrosion in pipes. While an increasing number of systems send such reports electronically at low cost, the requirement would increase costs for systems that still send reports by mail. Based on information from public water systems and state water agencies about the costs of complying with current requirements, CBO estimates that public water systems would spend about $14 million per year to comply with these requirements.

The bill would require public water systems that serve populations larger than 3,300 to conduct assessments of the risks posed to their systems by security threats and natural hazards and to prepare response plans. The bill would require those systems to certify to the EPA that they have conducted such assessments once every five years. Alternatively, systems could satisfy those requirements by certifying to the EPA that they are following consensus technical standards developed by the water industry and recognized by the EPA. Risk assessments are increasingly common in the water industry, and CBO expects that many systems, especially those that serve major populations, would already be in compliance because they follow industry standards; additional costs to them resulting from the mandate would be small. However, CBO expects that other systems, particularly those that are smaller in size, would need to conduct risk assessments and prepare response plans at varying costs, depending on their size and complexity. Based on information from the EPA and the American Water Works Association, CBO estimates that systems would spend an additional $25 million to comply with those requirements. That estimate is based on the expectation that many smaller systems would comply by conducting assessments at low cost using free assessment tools, while larger systems would undertake much more expensive and comprehensive analyses ranging into the hundreds of thousands of dollars.

The bill would impose a mandate by requiring the EPA to expand the number of small public water systems (those serving fewer than 10,000 people) that must monitor drinking water for unregulated contaminants. The EPA would select a representative sample of those systems to conduct monitoring. The bill would authorize the appropriation of $15 million per year to cover the costs of laboratory analysis of samples. However, systems would incur costs to collect samples and to train staff. Based on information from public water systems and state water agencies about the costs of sample collection under current requirements, CBO estimates that systems selected for monitoring would spend, in the aggregate, $2 million to $3 million each year to comply with those requirements.

Mandates on Public Entities

The bill would require SERCs and local emergency planning committees to notify state water agencies whenever there is a release of hazardous chemicals into water bodies used for drinking water and also would require water agencies to in turn notify public water systems in the affected area. Additionally, the bill would require SERCs and local emergency planning committees to provide information about chemicals stored at specific facilities whenever local public water systems request that information. Because it is already common practice for SERCs and local emergency committees to conduct such activities, CBO estimates that the costs of compliance would be small.

Other Effects on Public Entities

Under the bill, state and tribal agencies that have chosen to implement the Safe Drinking Water Act would likely incur additional costs to provide financial and technical assistance to public water systems that are subject to federal regulations under that act. Specifically, state and tribal water agencies would work with public water systems to meet requirements under the bill relating to consumer confidence reports, risk assessments, and monitoring for unregulated contaminants. Costs incurred by those agencies, however, would result from participation in a voluntary federal program.

The bill also includes a provision that would provide state and tribal governments with the authority to compel public water systems that are out-of-compliance with federal drinking water standards to undergo consolidation with another system, or to transfer ownership. In cases where the targeted systems are unable to meet federal drinking water standards, and are either financially unable or unwilling to take actions that would result in compliance, states and tribes with primary enforcement responsibility for the SDWA could require the owner or operator of such a system to assess options for consolidation or transfer and then carry out those actions if doing so is economically feasible and likely to result in greater compliance with federal standards. CBO expects that state and tribal agencies would generally use the authority selectively to focus on systems that have serious violations of drinking water standards; however, use of this authority could result in significant costs for some water systems, depending on how the authority is exercised, and the size and complexity of the systems affected. Because state and tribal water agencies would exercise the authority at their discretion, any costs incurred by affected water systems would not stem from a federal intergovernmental mandate under UMRA. Based on evidence from consolidation efforts in California and other states, CBO expects that many state and tribal agencies would provide financial assistance to cover necessary interconnection, improvement, and administrative costs for systems required to undergo consolidation or transfer. The bill also would authorize states and tribes to use federal funds provided through the DWSRF programs to cover the costs of consolidations and transfers.

Finally, the bill would benefit public water systems, as well as state, local, and tribal agencies that implement federal drinking water regulations, by authorizing federal financial and technical assistance for several drinking water grant programs. Public water systems would benefit from loans provided by state agencies for drinking water infrastructure projects. The bill would authorize the appropriation of $8 billion over the 2018–2022 period for the EPA to provide capitalization grants to DWSRFs to finance those loans. Any costs public entities might incur relating to grant and loan programs, including matching contributions, would result from conditions of federal assistance.

Estimate prepared by: Federal spending: Jon Sperl; Federal revenues: Staff of the Joint Committee on Taxation; Impact on state, local, and tribal governments: Jon Sperl; Impact on the private sector: Amy Petz.

Estimate approved by: H. Samuel Papenfuss, Deputy Assistant Director for Budget Analysis.

Federal Mandates Statement

The Committee adopts as its own the estimate of Federal mandates prepared by the Director of the Congressional Budget Office pursuant to section 423 of the Unfunded Mandates Reform Act.

Statement of General Performance Goals and Objectives

Pursuant to clause 3(c)(4) of rule XIII, the general performance goal or objective of this legislation is to amend the Safe Drinking Water Act to improve public water systems and enhance compliance with such Act.

Duplication of Federal Programs

Pursuant to clause 3(c)(5) of rule XIII, no provision of H.R. 3387 is known to be duplicative of another Federal program, including any program that was included in a report to Congress pursuant to section 21 of Public Law 111–139 or the most recent Catalog of Federal Domestic Assistance.

Committee Cost Estimate

Pursuant to clause 3(d)(1) of rule XIII, the Committee adopts as its own the cost estimate prepared by the Director of the Congressional Budget Office pursuant to section 402 of the Congressional Budget Act of 1974. At the time this report was filed, the estimate was not available.

Earmark, Limited Tax Benefits, and Limited Tariff Benefits

Pursuant to clause 9(e), 9(f), and 9(g) of rule XXI, the Committee finds that H.R. 3387 contains no earmarks, limited tax benefits, or limited tariff benefits.

DISCLOSURE OF DIRECTED RULE MAKINGS

Pursuant to section 3(i) of H. Res. 5, the following directed rule makings are contained in H.R. 3387:

- In section 2: Revision of regulations affecting consumer confidence reports under the amendment to SDWA section 1414(c)(4).
- In section 4: Promulgation of regulations implementing water system consolidation mandates under the amendment to SDWA section 1414(h).

ADVISORY COMMITTEE STATEMENT

No advisory committees within the meaning of section 5(b) of the Federal Advisory Committee Act were created by this legislation.

APPLICABILITY TO LEGISLATIVE BRANCH

The Committee finds that the legislation does not relate to the terms and conditions of employment or access to public services or accommodations within the meaning of section 102(b)(3) of the Congressional Accountability Act.

SECTION-BY-SECTION ANALYSIS OF THE LEGISLATION

Section 1. Short Title

Section 1 establishes the short title of the legislation as "the Drinking Water System Improvement Act of 2017".

Section 2. Improved Consumer Confidence Reports

Section 2 amends SDWA section 1414(c)(4) to institute certain changes related to requirements on the form, manner, and frequency that consumer confidence reports (CCR) are issued by community water systems.

First, section 2 amends SDWA section 1414(c)(4)(A) to permit CCRs to be mailed or provided by electronic means to drinking water system customers. The Committee understands that Americans are increasingly going away from a paper-driven society and instead relying on electronic technologies to access data, including real-time information. The Committee also recognizes that not all persons have access to or are comfortable using these means and intends that this new option not be used as an opportunity to avoid making paper copies available to those customers that want them.

Second, section 2 requires, under SDWA section 1414(c)(4)(B), three new types of information—relevant to the community water system's reporting period—that a system must report in its CCR. These include: (1) its compliance with corrosion control requirements, (2) identification, if any, of system-wide exceedances of the lead action level that required corrective action by EPA or a State exercising primary enforcement responsibility, and (3) an identification, if any, of SDWA violations that occurred.

Third, section 2 requires EPA, within 24 months of the date of enactment of the Drinking Water System Improvement Act, to revise the regulations implementing the CCR requirements for two issues: CCR content understandability and electronic delivery. Section 2 is intentionally narrow to these two issues to permit a targeted correction of these concerns and avoid reopening the entire rule. Specifically, in response to the December 2012 CCR Rule Retrospective Review, section 2 requires the rule revision to increase both the readability, clarity, and understandability of the information presented in the CCR as well as the accuracy and risk communication of the information presented. In addition, to reduce EPA's burden for issuing this rule revision, section 2 permits EPA to allow delivery of consumer confidence reports by methods consistent with methods described in the memorandum "Safe Drinking Water Act—Consumer Confidence Report Rule Delivery Options" issued by the Environmental Protection Agency on January 3, 2013.

Finally, section 2 requires as part of the rule revision to the requirements of SDWA section 1414(c)(4)(A) that community water systems serving 10,000 or more persons be obligated to provide, by mail, electronic means, or other methods permitted by the Administrator, a CCR to each customer of the system at least biannually. The Committee expects that when issuing these regulations, EPA will include feasible implementation options that reduce the burden on community water systems, States, and other relevant parties subject to the new requirements while maintaining the quality and availability of information for community water system customers.

Section 3. Contractual Agreements

During hearings by the Subcommittee on the Environment, the Subcommittee heard repeated testimony on the importance of encouraging partnerships between struggling public water systems and outside interests whose technical, financial, or managerial expertise would help that utility achieve compliance. Section 3 expands SDWA section 1414(h)(1) to permit an owner or operator of a public water system to enter into a contractual agreement for significant management or administrative functions of its public water system to correct its identified SDWA violations. The contract is intended to be part of a larger plan that is subject to approval by its State (if that State has primary enforcement responsibility for SDWA) or the EPA Administrator (if the State does not have primary enforcement responsibility). An approved plan would provide two years for the public water system to achieve compliance with its identified violations.

Section 3 also makes a technical change to correct the punctuation in SDWA section 1414(i)(1).

Section 4. Consolidation

Section 4(a) establishes new language at the end of SDWA section 1414(h) regarding transfers of ownership or consolidations that will help ensure safe drinking water. Specifically, the provisions are focused on the use of assessments, by public water systems whose produced drinking water creates a public health threat, to determine whether it makes sense for that utility to seek a transfer of ownership or consolidation with another utility.

In proposed SDWA section 1414(h)(3)(A), either a State with primary enforcement responsibility for SDWA or EPA, if the State does not have that authority under SDWA section 1413(a), may require the owner or operator of certain public water systems to assess their options for consolidation or transfer of ownership based on the presence of three conditions.

The first condition is that the public water system in question has repeatedly violated one or more SDWA requirements and this lack of compliance is likely to adversely affect human health. In addition, the public water system in question must have unsuccessfully tried to remedy these violations using technical assistance and a DWSRF loan or is unable or unwilling to undertake feasible and affordable actions suggested by either the State with primary enforcement or EPA to bring the water system into compliance.

In choosing the wording of the first prong of the first condition, the Committee does not intend any type of violation by a public water system to activate these provisions. Rather, the authority provided in proposed SDWA section 1414(h)(3) is limited to repeated and significant non-compliers whose systems are producing water that is unsafe for human consumption. The Committee does not intend this authority to be used for paperwork violations or when the water system is in significant compliance with SDWA requirements and is not producing water that threatens its customers' health.

The second condition is that a consolidation or transfer of the public water system is feasible, including feasibility based upon geographic considerations, technical concerns, access to capital, and chances for long-term success.

The last condition is that consolidation or transfer by the public water system could result in greater compliance with national primary drinking water regulations. The Committee intends that this condition be more than theoretical and incremental.

If all three conditions have been met and an assessment has been done, the State with primary enforcement responsibility or EPA, as appropriate, reviews the assessment. Upon completion of this review, under proposed SDWA section 1414(h)(3)(B), the State or EPA may require the owner or operator of the public water system to submit a plan for consolidation or transfer ownership. The plan's implementation becomes mandatory if three conditions are met: (1) the owner or operator of that water system has not taken steps to complete consolidation, not transferred ownership of the system, or could not achieve compliance after receiving technical assistance and a DWSRF loan; (2) after completing its assessment, the public water system violated another national primary drinking in a way that makes its produced water likely to adversely affect human health; and (3) the consolidation or transfer is feasible.

Section 4(a) also creates a new SDWA section 1414(h)(4) that permits, notwithstanding the limitation in SDWA section 1452(a)(3), DWSRF loans to be provided to public water systems trying to achieve compliance under this section through consolidation, transfer of ownership, or other means. SDWA section 1452(a)(3) generally prohibits the provision of DWSRF loans to public water systems in significant non-compliance or that that lack certain technical, managerial, or financial capabilities.

Finally, section 4(a) contains provisions extending legal protections for non-responsible parties consolidating with or acquiring ownership in a non-compliant public water system. In general, section 4(a) proposes a new SDWA section 1414(h)(5) that provides protection from any potential liability for damages arising from violations of the Safe Drinking Water Act that are identified in a plan for consolidation or ownership under SDWA section 1414(h)(3)(B).

To obtain this protection, the owner or operator of a public water system must take reasonable steps to identify, in the consolidation or ownership transfer plan they are submitting under proposed SDWA section 1414(h)(3), all potential violations of which they are aware and—as of the date the plan is submitted—any funds or other assets available to the public water system that committed such violation to satisfy its liability.

If, as appropriate, the State or the Administrator approves the public water system's plan for consolidation or transfer of ownership, under proposed SDWA section 1414(h)(5)(B), the public water system is not liable for a violation of the Safe Drinking Water Act identified in its plan, except to the extent to which funds or other assets have been identified in its plan to satisfy that liability.

Section 4 contains three other features. First, it requires, in proposed SDWA section 1414(h)(6), regulations to implement the consolidation or ownership transfer provisions within proposed paragraphs (3) through (5) to SDWA section 1414(h). Second, section 4(b) adds a new SDWA section 1413(a)(6) to require that States, as a condition of the primary enforcement delegation, have adopted and are implementing procedures consistent with the provisions in proposed paragraphs (3) through (6) to SDWA section 1414(h). Last, it makes a conforming amendment to SDWA section 1413(b), relating to EPA providing written notification to States about determinations of the primary enforcement authority status, to account for the proposed change to SDWA section 1413(a).

Section 5. Improved Accuracy and Availability of Compliance Monitoring Data

Section 5 requires EPA, in coordination with the State, public water systems, and other interested stakeholders to create a strategic plan for improving the accuracy and availability of monitoring data collected to demonstrate SDWA compliance, particularly data submitted by public water systems to States and data submitted by States to EPA. This strategic plan, including a summary of its findings and practicable and cost-effective recommendations to improve accuracy and availability of monitoring data collected to demonstrate SDWA compliance, is due to Congress not later than 1 year after the date of enactment of the Drinking Water System Improvement Act.

Due to software compatibility, budgeting, and management issues the Agency and States have faced in other electronic reporting programs, like the electronic manifest program for hazardous water under section 3024 of the Solid Waste Disposal Act (42 U.S.C. 6969g), the Committee is reluctant to require a solution without EPA working out potential issues on the front end. In developing the strategic plan and its recommendations under section 4, the Administrator is obligated to evaluate any challenges: (1) faced by States and public water systems in using electronic systems for data collection and dissemination, (2) in ensuring the accuracy and integrity of submitted data, and (3) regarding access to information and the usability of an electronic system. The Committee hopes the Administrator will take advantage of the authorities in section 5 that permit consultation with States and other Federal agencies that have experience using these kinds of systems.

Section 6. Asset Management

Section 6 amends SDWA section 1420 in three places to encourage the use of asset management by drinking water delivery systems.

The Subcommittee on the Environment received testimony about the importance of asset management in helping drinking water systems become economically sustainable. At the same time, witnesses stated that it was better to encourage this practice rather than mandate its use. The Committee believes technical assistance in this area, especially for smaller and rural systems, will be the most beneficial to seeing wider deployment of these practices.

First, section 6 requires States, as part of the Capacity Development Strategy, to consider, solicit, and include as appropriate, how the State will encourage the use of asset management plans and assist, including technical assistance, in the use of asset management best practices by public water systems as part of these plans. The Committee envisions that States, as appropriate, will revise their Capacity Development Strategies to incorporate this new information.

Second, section 6 requires that when the State publishes its Capacity Development Strategy report to detail the efficacy of and progress made on the State's efforts to encourage development of asset management plans and engage of relevant training to implement asset management plans.

Last, section 6 requires the Administrator to, every five years, review and update, if appropriate, educational materials made available by the Agency to owners, managers, and operators of public water systems, local officials, technical assistance providers (including non-profit water associations), and State personnel concerning best practices for asset management strategies that may be used.

Section 7. Community Water System Risk and Resilience

Using much of the architecture and policy objectives contained in existing SDWA section 1433, section 6(a) replaces the provisions in SDWA section 1433 regarding the creation of risk and resilience assessments and emergency response plans by community water systems serving more than 3,300 persons.

Proposed SDWA section 1433(a)(1) requires community water systems serving over 3,300 persons to assess the risks to, and resilience of, their system. Proposed SDWA section 1433(a)(1)(A) mandates that this assessment include a review of six elements. These include: (1) the CWS's risk from malevolent acts and natural hazards; (2) the resilience of the pipes and constructed conveyances, physical barriers, source water, water collection and intake, pretreatment, treatment, storage and distribution facilities, electronic, computer, or other automated systems (including the cyber security of such systems) utilized by the CWS; (3) the community water system's monitoring practices; (4) the financial infrastructure of the community water system, including cyber protections, for infrastructure of the CWS; (5) the CWS's use, storage, or handling of various chemicals; and (6) the CWS's operation and maintenance. In addition, proposed SDWA section 1433(a)(1)(B) permits the CWS to include in its assessment an evaluation of capital and operational needs for its risk and resilience management.

The Committee notes that proposed section 1433 uses the term "malevolent act" in place of the terms "terrorist attack or other intentional acts." The Committee does not intend the switching of these terms to be interpreted to mean that these activities are no longer covered. Rather, the Committee used the term "malevolent acts" to capture the term used in the drinking water utility sector to encompass the range of threats facings CWSs. When it comes to matters surrounding acts meant to substantially disrupt the ability of the system to provide a safe and reliable supply of drinking water, the Committee wants no ambiguity within the CWS sector about what types of assessments or responses need to be made.

The Committee also defined and included resilience to natural hazards in this section because the Committee understands that the drinking water utility sector currently assesses and addresses their risks holistically, looking at both intentional acts and natural hazards such as extreme weather. In the wake of Hurricanes Harvey, Irma, and Maria, resilience efforts like this by drinking water systems can prepare for and mitigate help impacts from extreme weather.

To aid community water systems in assessing potential risks from malevolent acts, proposed SDWA section 1433(a)(2) requires EPA, after consultation with appropriate Federal, State, and local departments and agencies and not later than August 1, 2019, to provide relevant baseline information to community water systems on malevolent acts that may substantially disrupt the ability of the CWS to provide a safe and reliable supply of drinking water or might otherwise present significant public health or economic concerns to the community served by the CWS.

Proposed SDWA section 1433(a)(3)(A) requires community water systems serving more than more 3,300 persons to submit a certification to EPA that the CWS has completed the assessment it is mandated to do under proposed SDWA section 1433(a)(1). As is done in existing SDWA section 1433(a)(2), proposed SDWA section 1433(a)(3)(A) creates a staggered deadline for submission of the required certification based on the size of the CWS. Specifically, CWSs serving a population of 100,000 or more persons must submit their certification by March 31, 2020; CWSs serving a population of between 50,000 and 99,999 persons must submit their certification by December 31, 2020; and CWSs serving a population between 3,301 persons and 49,999 must submit their certification by June 30, 2021.

Of note, proposed SDWA section 1433(a)(3)(B) requires each CWS that performed a risk and resilience assessment to review their assessment every 5 years—from the date that its certification was due to EPA under proposed section 1433(a)(3)(A)—to determine whether its assessment needs to be revised. Similar to proposed SDWA section 1433(a)(3)(A), once the CWS completes its quintennial review of its assessment, proposed SDWA section 1433(a)(3)(B) requires the CWS to submit a certification to EPA that it has reviewed its assessment and, if applicable, revised its risk and resilience assessment.

Proposed SDWA section 1433(a)(4) details the contents required to be made a part of the certification. This section states that the certifications are limited to three pieces of information: (1) the identity of the community water system submitting the certification,

(2) the date of the certification, and (3) a statement that the community water system has conducted, reviewed, or revised the assessment. Since proposed SDWA section 1433 removes the broad information protections in existing SDWA section 1433, any information protection afforded to a CWS will come from EPA's determination that it meets the criteria of the Freedom of Information Act (FOIA). Any CWS unsure whether submitting additional information on their certification will be protected by FOIA, submits that data at it and its customers' risk.

Proposed SDWA section 1433(a)(5) also retains existing provisions in SDWA section 1433 that prevent a CWS from being required to provide, under State or local law, a risk and resilience assessment (or any revision of it) to any State, regional, or local governmental entity solely because proposed SDWA section 1433 required the CWS to submit a certification to EPA.

The Committee believes the use of certifications in the newly proposed SDWA section 1433 represents an important compromise from existing law and one that addresses concerns about protecting very sensitive information from public disclosure, while at the same time removing obligations on EPA that made use of the information difficult and erected substantial financial, storage, and disposal challenges for the EPA.

Section 7(a) also retains the existing SDWA requirement for emergency response plans. Proposed SDWA section 1433(b) requires a CWS serving more than 3,300 people to prepare or revise an emergency response plan that incorporates findings of its risk and resilience assessment under proposed SDWA section 1433(a)(1). In the same form and manner as established under proposed SDWA section 1433(a)(3), proposed SDWA section 1433(b)(1) requires each CWS to certify to EPA, as soon as reasonably possible after the date of enactment of the Drinking Water System Improvement Act of 2017, but not later than six months after completion of its risk and resilience assessment, that the CWS complete its emergency response plan. Proposed SDWA section 1433(b) requires four elements be included in the emergency response plan: (1) strategies and resources to improve the resilience of the CWS, including the physical and cyber security of the CWS; (2) implementable plans and procedures and identification of equipment that can be utilized in the event of a malevolent act or natural hazard that threatens the ability of the CWS to deliver safe drinking water; (3) actions, procedures, and equipment that can obviate or significantly lessen the impact of a malevolent act or natural hazards on public health and the supply of drinking water; and (4) usable strategies to aid in the detection of malevolent acts or natural hazards that threaten the security or resilience of the CWS.

Proposed SDWA section 1433(c) is designed to ensure a harmonized response to actual events at the CWS. Specifically, the CWSs is mandated, to the extent possible, to coordinate with existing local emergency planning committees established under the Emergency Planning and Community Right-To-Know Act of 1986 when preparing or revising a risk and resilience assessment or emergency response plan.

Newly proposed SDWA section 1433(d), which requires each CWS required to do a risk and resilience assessment or emergency response plan under section 1433 to maintain, for five years, a copy of the assessment and the emergency response plan (including any revised assessment or plan) submitted to EPA.

Proposed SDWA section 1433(e) requires EPA to provide guidance and technical assistance to community water systems serving a population of less than 3,300 persons on how to conduct resilience assessments, prepare emergency response plans, and address threats from malevolent acts and natural hazards that threaten to disrupt the provision of safe drinking water or significantly affect the public health or the safety or supply of drinking water. Even though these sized water systems are not required to do a risk and resilience assessment or devise an emergency response plan, the Committee notes the value that programs like this provided to smaller utilities trying to protect themselves from malevolent acts.

Proposed SDWA section 1433(f) creates a path to compliance with this section that is separate from the requirements contained in proposed subsections (a)(1) and (b) of SDWA section 1433. Specifically, proposed SDWA section 1433(f)(1) permits a CWS to meet some of its compliance obligations under this section by using and complying with technical standards that the EPA Administrator, pursuant to section 12(d) of the National Technology Transfer and Advancement Act of 1995, has recognized as a means of satisfying the requirements proposed in sections 1433(a)(1) and 1433(b). Regardless of its use of proposed section 1433(f) to satisfy compliance with risk and resilience assessment and emergency response plans, the CWS is still obligated to submit certifications to EPA acknowledging completion of the assessment and plan requirements.

The Committee believes the alternate compliance path provided in proposed SDWA section 1433(f) will increase overall compliance by CWSs and reduce the administrative burden on EPA and regulated stakeholders. The Committee recognizes that, because of changes in technology and emerging threats, water systems may be reluctant to make upgrades if they are concerned with ensuring regulatory compliance. The Committee's language intends to capitalize on efforts that have been organically occurring over the last 10 years in the drinking water sector to improve detection and response to terrorism and other natural disasters. The Committee has taken specific notice of efforts taken by drinking water utilities, such as J–100, the American Water Work Association/American National Standards Institute voluntary consensus standard encompassing an all-hazards risk and resilience management process, and EPA's workshops and tabletop exercises for terrorism and other national hazards facing drinking water systems in smaller and rural communities. For this reason, the language of proposed section 1433(f)(2) deploys the Federal government's existing practice, under section 12(d) of the National Technology Transfer and Advancement Act of 1995, of recognizing technical standards developed or adopted by third-party organizations or voluntary consensus standards bodies that carry out the policy objectives of or activities required by Federal law.

Proposed SDWA section 1433(g) creates an EPA program, called the Drinking Water Infrastructure Risk and Resilience Program, under which EPA can award grants to owners or operators of CWSs to help increase their resilience. Grants made available under this subsection are authorized for five years, from fiscal year 2018 through 2022.

Under proposed SDWA section 1433(g)(2), an owner or operator of a CWS receiving a grant from the Drinking Water Infrastructure Risk and Resilience Program is required to use the grant funds exclusively to assist in the planning, design, construction, or implementation of a program or project consistent with its emergency response plan. Eligible expenses include a range of items and activities, including the purchase and installation of equipment to detect drinking water contaminants or malevolent acts; fencing, gating, lighting, or security cameras; treatment technologies and equipment to improve drinking water system resilience; improvements to electronic, computer, financial, or other automated systems and remote systems; and participation in training programs, and the purchase of training manuals and guidance materials, relating to security and resilience.

Proposed SDWA section 1433(g)(3) precludes grants awarded from the Drinking Water Infrastructure Risk and Resilience Program from being used for used for personnel costs, or for monitoring, operation, or maintenance of facilities, equipment, or systems.

Proposed SDWA section 1433(h)(4) authorizes, for fiscal years 2018 through 2022, $5,000,000 from the funds provided to the Drinking Water Infrastructure Risk and Resilience Program to provide technical assistance to community water systems to assist in responding to and alleviating a vulnerability that would substantially disrupt the CWS system from providing a safe and reliable supply of drinking water that EPA determines to present an immediate and urgent need.

Proposed SDWA section 1433(h)(5) authorizes, for fiscal years 2018 through 2022, $10,000,000 from the funds provided to the Drinking Water Infrastructure Risk and Resilience Program to provide grants to community water systems serving a population of less than 3,300 persons, or nonprofit organizations receiving assistance under SDWA section 1442(e), for technical assistance activities that are consistent with proposed SDWA section 1433(e).

Proposed SDWA section 1433(h)(6) authorizes appropriations of $35,000,000 for each of fiscal years 2018 to 2022 to carry out SDWA section 1433(h).

Finally, proposed section 1433(h) creates the operable definitions for proposed for use in SDWA section 1433.

The term "resilience" means the ability of a community water system or an asset of a community water system to adapt to or withstand the effects of a malevolent act or natural hazard without interruption to the asset's or system's function, or if the function is interrupted, to rapidly return to a normal operating condition.

The term "natural hazard" means a natural event that threatens the functioning of a community water system, including an earthquake, tornado, flood, hurricane, wildfire, and hydrologic changes.

Section 7(b)(1) makes a clarification regarding the treatment of information obtained by EPA under SDWA section 1433 prior to the date of enactment the Drinking Water System Improvement Act. The Committee understands that even though many years have passed since community water systems submitted their vulnerability information to EPA, even an accidental release of information could compromise the security of America's water system or the public's health. For this reason, section 7(b) retains, for information previously submitted to EPA pursuant to SDWA section 1433, the protections from public disclosure that were in effect on the day before enactment of the Drinking Water System Improvement Act.

Section 7(b)(2) attempts to address whether and how to dispose of vulnerability assessments and related information obtained by EPA more than 10 years ago. Due to stringent statutory requirements on access controls, handling, and providing this sensitive information submitted to others, EPA has been unable to find a cost-effective way to return potentially outdated vulnerability assessments to the submitting CWS without risking unauthorized disclosure or otherwise violating the law. Section 7(b)(2) requires EPA, in partnership with community water systems to develop a strategy to, timely, securely, and permanently dispose of, or return to the applicable community water system, any information EPA obtained from it that contains protected information.

Section 8. Authorization for Grants for State Programs

Section 8 reauthorizes appropriations for the Public Water System Supervision grants under SDWA section 1443(a)(7) at $150,000,000 in each of fiscal years 2018 through 2022.

Section 9. Monitoring for Unregulated Contaminants

Section 9 amends SDWA section 1445 to require, no earlier than three years after the date of enactment of the Drinking Water System Improvement Act and subject to certain conditions, monitoring of unregulated contaminants by public water systems serving between 3,300 and 10,000 persons.

Specifically, section 9(a) suspends the SDWA section 1445(a)(2)(A) requirement to only require a representative sample of public water systems serving 10,000 persons or fewer to monitor for unregulated contaminants. In its place, section 9(a) requires, subject to the availability of appropriations and a determination of sufficient laboratory capacity to accommodate the additional analyses, that public water systems serving between 3,300 and 10,000 persons be subject to monitoring requirements and that a system serving less than 3,300 be subject to monitoring only as part of a representative sample.

Finally, section 9(c) amends SDWA section 1445(g)(7) to require unregulated contaminant monitoring data collected using this new, broader universe of systems serving between 3,300 and 10,000 persons, be included in EPA's Unregulated Contaminants Data Base unless it duplicates monitoring information obtained from similarly sized public water systems.

The Committee is aware of the compliance burden that new monitoring could create for many smaller public water systems, especially since utilities subject to monitoring as part of a representative sample have their mailing and testing costs paid for by the Federal government. The Committee took care to protect against this burden in two ways. First, section 9(b) does not affect the intent and operation of SDWA sections 1445(a)(2)(H) and 1452(o)— but extends the authorization in SDWA section 1445(a)(2)(H) through fiscal year 2022—and section 9(a) adds an additional authorization of $15,000,000 in appropriations for this purpose. Second, to avoid systems' non-compliance due to their inability to afford it without Federal aid, the Committee conditioned the requirement on EPA, in proposed SDWA section 1445(j)(1), to require monitoring for systems serving between 3,300 and 10,000 on the availability of appropriations. If the appropriations are not available to address the burdens to EPA and water systems, the Committee intends that EPA revert to its existing practice of using a representative sample for systems serving a population of 10,000 or fewer.

The Committee supports EPA's use of high quality science in its work, but is concerned that invalid, incomplete, or incorrectly gathered monitoring data will compromise its value to EPA, particularly in terms of meeting EPA's statistical modeling and analysis of it. Since the Committee does not wish to place water systems in the "Catch-22" position of complying with the requirement to obtain the sample, but unable to have an approved laboratory to analyze the sample; section 9(a) contains proposed SDWA section 1445(j)(3), which permits EPA to stop these new monitoring requirements if there is not sufficient laboratory capacity to carry out the sample analysis. Should EPA determine there is not sufficient laboratory capacity to handle the increase in required monitoring, the Committee intends that EPA revert to its existing practice of using a representative sample for systems serving a population of 10,000 or fewer.

Section 10. State Revolving Loan Funds

Section 10 makes different amendments to provisions related to State Revolving Loan Funds under SDWA section 1452.

Section 10(a) clarifies and expands the types of eligible expenditures permitted from a State DWSRF under SDWA section 1452(a)(2)(B), conditioned on the EPA Administrator determining that they facilitate compliance with national primary drinking water standards or significantly further the public health objectives of the SDWA. Specifically, "siting" is now its own expenditure—and not a subset of "an associated preconstruction activity"—and "replacing or rehabilitating aging treatment, storage, or distribution facilities of public water systems" becomes an explicit eligible use.

Section 10(b) also amends the American Iron and Steel Products purchase requirements in SDWA section 1452(a)(4) by extending its application to fiscal years 2018 through 2022.

Section 10(c) adds a new SDWA section 1452(a)(5) that requires, between fiscal years 2018 and 2022, water utilities serving a population of more than 10,000 can only obtain DWSRF funding if they considered the cost and effectiveness of the relevant processes, materials, techniques, and technologies for carrying out their project and certified this consideration to its State.

This provision is not meant to convey a preference for any materials nor to make cost the sole feature of any consideration. Rather, this language is an effort to ensure DWSRF money is going to projects where recipients have considered both the cost as well as the effectiveness of the relevant processes, materials, techniques, and technologies that public money is purchasing.

The Committee is aware that section 602(b)(13) of the Federal Water Pollution Control Act (Clean Water Act) currently contains a related mandate on recipients of Clean Water Act State Revolving Funds that could serve as a model. Rather than using the exact same mandate in the Safe Drinking Water Act, the Committee was worried about the potential burden this could place on States and those receiving assistance from the DWSRF. The Committee consciously chose to use a less prescriptive and burdensome version of Clean Water Act section 602(b)(13).

Whereas the Clean Water Act language requires a waste water utility to "study and evaluate" the cost and effectiveness of the processes, materials, techniques, and technologies used in the project funded by the SRF money, section 10(c) of the Drinking Water System Improvement Act only requires the system to "consider" them. The Committee does not intend this consideration to be a perfunctory exercise, an endless analysis by the community, or litigated by goods or services providers that are not selected for a project. Consideration is satisfied when a community's relevant authority thinks about reasonably available, technically and economically feasible products to complete the project, what their respective costs might be, and whether the technical or environmental conditions merit their use.

Importantly, the required consideration is not binding on the decision made by a community. Communities have several reasons for making the decisions that they do and the Committee believes those are discussions that need to occur between community decision makers and the users of the system.

The Clean Water Act language places requirements on the selection of projects by SRF applicants as well as certain needs that must be achieved through the selection process. The language in section 10(c), however, intentionally omits selection criteria mandates. The Committee is concerned that doing so would sever the important relationship between local communities and their engineer of record in designing a water project in the manner that best serves the unique needs and considerations of local communities.

Finally, section 10(c), like the Clean Water Act, requires a certification to the State—presumably the office responsible for operating its DWSRF—that the consideration has been made by the DWSRF applicant. The Committee does not intend this to be a burdensome or involved process and is concerned this requirement could be misunderstood and develop into an onerous approval process, which could unnecessarily delay projects. The Committee understands that some States require only a signed check list to demonstrate compliance to the State under the Clean Water Act; that simple arrangement would more than meet the Committee's expectation for fulfillment of the certification.

Section 10(d) institutes a new SDWA section 1452(a)(6) that applies Federal requirements regarding prevailing wage treatment for laborers and mechanics to drinking water loan funded construction under the Safe Drinking Water Act. The practical value of this provision is that it locates the prevailing wage requirement in the SDWA. Congress already has applied Davis-Bacon prevailing wage requirements through appropriations law to DWSRF program funding for fiscal year 2012 and all future years.

Section 10(e) makes two changes to SDWA section 1452(d)(2) related to the amount of loan subsidies made available to disadvantaged communities by a State DWSRF in a fiscal year, including those communities a State expects to become disadvantaged because of its proposed project.

Under section 10(e), the ceiling on the amount of DWSRF assistance used for these purposes is raised from 30 percent to 35 percent of the State's capitalization grant.

Additionally, section 10(e) institutes a minimum requirement that 6 percent of a State's annual DWSRF capitalization grant be dedicated to loan subsidies made available to eligible water systems meeting the definition of a disadvantaged community under SDWA section 1452(d)(3). If a State does not have enough applications for DWSRF assistance from eligible disadvantaged communities that total 6 percent of the State's annual DWSRF capitalization grant, the State may use these funds for other worthy DWSRF applicants.

Section 10(f) makes changes to SDWA section 1452(f)(1) regarding repayment of principal and interest on loans issued by a State DWSRF. Existing law provides that principal and interest payments begin no later than one year after completion of the project; that each loan to be fully amortized within 20 years of the project's completion; and that States may provide disadvantaged communities an extended loan period of 30 years after the date of project's completion so long as the extended term does not exceed the expected design life of the project.

Under section 10(f), principal and interest payments cannot begin later than 18 months after completion of the project on which the loan was made; requires each loan to be fully amortized within 30 years of the project's completion; and permits States to allow disadvantaged communities an extended loan period of 40 years after the date of project's completion so long as the extended term does not exceed the expected design life of the project.

Section 10(g) amends SDWA section 1452(h) to require EPA, in any needs assessment after the date of enactment of the Drinking Water System Improvement Act, to include an assessment of costs to replace all lead service lines (as defined in SDWA section 1459B(a)(4)) of all eligible public water systems in the United States. To help provide a more granular picture of lead service lines, section 10(g) requires EPA's assessment to separately describe the costs associated with replacing the portions of lead service lines that are owned by an eligible public water system and, to the extent practicable, the costs associated with replacing any remaining portions of lead service lines, including those owned by private residences.

The Committee wishes to note here that it is aware of ongoing legal questions related to the ownership of lead service lines. The Committee does not wish to use this bill to take a position on that question and intentionally drafted to avoid any implication that Congress was taking an opinion on this matter.

Section 10(h) removes the current restriction in SDWA section 1452(k)(1)(C) on States using a portion of their DWSRF capitalization grant to delineate and assess source water protection areas in accordance with SDWA section 1453. Section 10(h) retains the requirement in SDWA section 1452(k)(1)(C) that funds set aside for this purpose be obligated within four fiscal years.

Section 10(i) reauthorizes appropriations to carry out SDWA section 1452 and operation of the DWSRF, providing $8 million over five years. Specifically, section 10(i) authorizes $1.2 billion in fiscal year 2018, $1.4 billion in fiscal year 2019, $1.6 billion in fiscal year 2020, $1.8 billion in fiscal year 2021, and $2 billion in fiscal year 2022.

Section 10(j) creates a new SDWA section 1452(s) related to best practices for the DWSRF. Specifically, EPA is authorized to collect—within three years—information from States on efforts and practices related to streamlining and aiding the DWSRF application process; spending of DWSRF funds and types of assistance granted; and enhancing management of and use of key financial measures for their DWSRFs. EPA is then required to take this information and make publicly available those best practices from among the data it has collected.

Section 11. Authorization for Source Water Petition Programs

Section 11 extends the reauthorization of appropriations from fiscal years 2018 through 2022 to carry out the Source Water Petition Program under SDWA section 1454(e). The program provides grants to States that establish a voluntary source water protection partnership program that meets EPA guidelines and is approved by the Administrator.

Section 12. Review of Technologies

Section 12 creates a new SDWA section 1459C dedicated to innovative efforts to protect public health. This section authorizes $10 million for EPA to review existing and potential methods, means, equipment, and intelligent systems or other smart smart-technology to: (1) ensure the physical integrity of a community water system; (2) prevent, detect, or respond in real-time to regulated contaminants in drinking water and source water; (3) allows for use of alternate drinking water supplies from non-traditional sources; and (4) facilitate source water assessments and protection.

Section 13. Drinking Water Fountain Replacement for Schools

Section 13 creates a new SDWA section 1465. Under this section, EPA is required to establish a grant program, authorized at $5,000,000 per year for fiscal years 2018 through 2022, to provide assistance to schools and daycare centers containing drinking water fountains manufactured before 1988. Specifically, the grants are to be used to replace those drinking water fountains and may be used to pay for monitoring lead levels in those schools or daycare centers.

Section 13 also requires that priority for awarding these grants should go to schools and daycare centers based upon economic need. The Committee intends "economic need" to be interpreted to mean that the school or daycare would otherwise have trouble obtaining the resources to make this improvement.

Section 14. Source Water

Section 14 amends the Emergency Planning and Community Right to Know Act (EPCRA) to help community water systems better understand real and potential threats to the source water they treat for drinking water.

First, section 14 amends EPCRA section 304 to have a State emergency response commission notify the State office primarily responsible for drinking water if a regulated entity has an unauthorized release to the source water of a community water system. Once notified, the State office primarily responsible for drinking water then alerts any community water system whose source water is affected by such release.

In addition, section 14 amends EPCRA section 312(e) to permit community water systems to have access to information on the types of hazardous chemicals located at facilities near the source water they use for drinking water.

Section 15. Report on Federal cross-cutting requirements

The Subcommittee on the Environment received testimony on the impact of cross-cutting requirements. Section 15 requires the Government Accountability Office, within one year of the date of enactment of the Drinking Water System Improvement Act, to conduct a study and issue a report to Congress that identifies demonstrations of compliance with a State or local environmental law that may be substantially equivalent to any demonstration required by the Administrator for compliance with a Federal cross-cutting requirement (a requirement that is a condition for receipt of Federal funding). The study is supposed to be conducted in consultation with EPA, State agencies that have primary enforcement responsibility for SDWA, and public water systems.

Changes in Existing Law Made by the Bill, as Reported

In compliance with clause 3(e) of rule XIII of the Rules of the House of Representatives, changes in existing law made by the bill, as reported, are shown as follows (existing law proposed to be omitted is enclosed in black brackets, new matter is printed in italic, and existing law in which no change is proposed is shown in roman):

Safe Drinking Water Act:
Title XIV—Safety Of Public Water Systems

Part B—Public Water Systems

State Primary Enforcement Responsibility

Sec. 1413.

(a) For purposes of this title, a State has primary enforcement responsibility for public water systems during any period for which the Administrator determines (pursuant to regulations prescribed under subsection (b)) that such State—

(1) has adopted drinking water regulations that are no less stringent than the national primary drinking water regulations promulgated by the Administrator under subsections (a) and (b) of section 1412 not later than 2 years after the date on which the regulations are promulgated by the Administrator, except that the Administrator may provide for an extension of not more than 2 years if, after submission and review of appropriate, adequate documentation from the State, the Administrator determines that the extension is necessary and justified;

(2) has adopted and is implementing adequate procedures for the enforcement of such State regulations, including conducting such monitoring and making such inspections as the Administrator may require by regulation;

(3) will keep such records and make such reports with respect to its activities under paragraphs (1) and (2) as the Administrator may require by regulation;

(4) if it permits variances or exemptions, or both, from the requirements of its drinking water regulations which meet the requirements of paragraph (1), permits such variances and exemptions under conditions and in a manner which is not less stringent than the conditions under, and the manner in, which variances and exemptions may be granted under sections 1415 and 1416;

(5) has adopted and can implement an adequate plan for the provision of safe drinking water under emergency circumstances including earthquakes, floods, hurricanes, and other natural disasters, as appropriate [; and];

(6) has adopted and is implementing procedures for requiring public water systems to assess options for, and complete, consolidation or transfer of ownership, in accordance with the regulations issued by the Administrator under section 1414(h)(6); and

[(6)] *(7)* has adopted authority for administrative penalties (unless the constitution of the State prohibits the adoption of the authority) in a maximum amount—

(A) in the case of a system serving a population of more than 10,000, that is not less than $1,000 per day per violation; and

(B) in the case of any other system, that is adequate to ensure compliance (as determined by the State); except that a State may establish a maximum limitation on the total amount of administrative penalties that may be imposed on a public water system per violation.

(b) (1) The Administrator shall, by regulation (proposed within 180 days of the date of the enactment of this title), prescribe the manner in which a State may apply to the Administrator for a determination that the requirements [of paragraphs (1), (2), (3), and (4)] of subsection (a) are satisfied with respect to the State, the manner in which the determination is made, the period for which the determination will be effective, and the manner in which the Administrator may determine that such requirements are no longer met. Such regulations shall require that before a determination of the Administrator that such requirements are met or are no longer met with respect to a State may become effective, the Administrator shall notify such State of the determination and the reasons therefor and shall provide an opportunity for public hearing on the determination. Such regulations shall be promulgated (with such modifications as the Administrator deems appropriate) within 90 days of the publication of the proposed regulations in the Federal Register. The Administrator shall promptly notify in writing the chief executive officer of each State of the promulgation of regulations under this paragraph. Such notice shall contain a copy of the regulations and shall specify a State's authority under this title when it is determined to have primary enforcement responsibility for public water systems.

(2) When an application is submitted in accordance with the Administrator's regulations under paragraph (1), the Administrator shall within 90 days of the date on which such application is submitted (A) make the determination applied for, or (B) deny the application and notify the applicant in writing of the reasons for his denial.

(c) Interim Primary Enforcement Authority.—A State that has primary enforcement authority under this section with respect to each existing national primary drinking water regulation shall be considered to have primary enforcement authority with respect to each new or revised national primary drinking water regulation during the period beginning on the effective date of a regulation adopted and submitted by the State with respect to the new or revised national primary drinking water regulation in accordance with subsection (b)(1) and ending at such time as the Administrator makes a determination under subsection (b)(2)(B) with respect to the regulation.

Enforcement of Drinking Water Regulations

Sec. 1414.

(a)

(1)

(A) Whenever the Administrator finds during a period during which a State has primary enforcement responsibility for public water systems (within the meaning of section 1413(a)) that any public water system—

(i) for which a variance under section 1415 or an exemption under section 1416 is not in effect, does not comply with any applicable requirement, or

(ii) for which a variance under section 1415 or an exemption under section 1416 is in effect, does not comply with any schedule or other requirement imposed pursuant thereto, he shall so notify the State and such public water system and provide such advice and technical assistance to such State and public water system as may be appropriate to bring the system into compliance with the requirement by the earliest feasible time.

(B) If, beyond the thirtieth day after the Administrator's notification under subparagraph (A), the State has not commenced appropriate enforcement action, the Administrator shall issue an order under subsection (g) requiring the public water system to comply with such applicable requirement or the Administrator shall commence a civil action under subsection (b).

(2) Enforcement in Nonprimacy States.—

(A) In General.—If, on the basis of information available to the Administrator, the Administrator finds, with respect to a period in which a State does not have primary enforcement responsibility for public water systems, that a public water system in the State—

(i) for which a variance under section 1415 or an exemption under section 1416 is not in effect, does not comply with any applicable requirement; or

(ii) for which a variance under section 1415 or an exemption under section 1416 is in effect, does not comply with any schedule or other requirement imposed pursuant to the variance or exemption; the Administrator shall issue an order under subsection (g) requiring the public water system to comply with the requirement, or commence a civil action under subsection (b).

(B) Notice.—If the Administrator takes any action pursuant to this paragraph, the Administrator shall notify an appropriate local elected official, if any, with jurisdiction over the public water system of the action prior to the time that the action is taken.

(b) The Administrator may bring a civil action in the appropriate United States district court to require compliance with any applicable requirement, with an order issued under subsection (g), or with any schedule or other requirement imposed pursuant to a variance or exemption granted under section 1415 or 1416 if—

(1) authorized under paragraph (1) or (2) of subsection (a), or

(2) if requested by (A) the chief executive officer of the State in which is located the public water system which is not in compliance with such regulation or requirement, or (B) the agency of such State which has jurisdiction over compliance by public water systems in the State with national primary drinking water regulations or State drinking water regulations.

The court may enter, in an action brought under this subsection, such judgment as protection of public health may require, taking into consideration the time necessary to comply and the availability of alternative water supplies; and, if the court determines that there has been a violation of the regulation or schedule or other requirement with respect to which the action was brought, the court may, taking into account the seriousness of the violation, the population at risk, and other appropriate factors, impose on the violator a civil penalty of not to exceed $25,000 for each day in which such violation occurs.

(c) Notice to States, the Administrator, and Persons Served.—

(1) In General.—Each owner or operator of a public water system shall give notice of each of the following to the persons served by the system:

(A) Notice of any failure on the part of the public water system to—

(i) comply with an applicable maximum contaminant level or treatment technique requirement of, or a testing procedure prescribed by, a national primary drinking water regulation; or

(ii) perform monitoring required by section 1445(a).

(B) If the public water system is subject to a variance granted under subsection (a)(1)(A), (a)(2), or (e) of section 1415 for an inability to meet a maximum contaminant level requirement or is subject to an exemption granted under section 1416, notice of—

(i) the existence of the variance or exemption; and

(ii) any failure to comply with the requirements of any schedule prescribed pursuant to the variance or exemption.

(C) Notice of the concentration level of any unregulated contaminant for which the Administrator has required public notice pursuant to paragraph (2)(F).

(D) Notice that the public water system exceeded the lead action level under section 141.80(c) of title 40, Code of Federal Regulations (or a prescribed level of lead that the Administrator establishes for public education or notification in a successor regulation promulgated pursuant to section 1412).

(2) Form, Manner, and Frequency Of Notice.—

(A) In General.—The Administrator shall, by regulation, and after consultation with the States, prescribe the manner, frequency, form, and content for giving notice under this subsection. The regulations shall—

(i) provide for different frequencies of notice based on the differences between violations that are intermittent or infrequent and violations that are continuous or frequent; and

(ii) take into account the seriousness of any potential adverse health effects that may be involved.

(B) State Requirements.—

(i) In General.—A State may, by rule, establish alternative notification requirements—

(I) with respect to the form and content of notice given under and in a manner in accordance with subparagraph (C); and

(II) with respect to the form and content of notice given under subparagraph (E).

(ii) Contents.—The alternative requirements shall provide the same type and amount of information as required pursuant to this subsection and regulations issued under subparagraph (A).

(iii) Relationship to Section 1413.—Nothing in this subparagraph shall be construed or applied to modify the requirements of section 1413.

(C) Notice of Violations or Exceedances With Potential to Have Serious Adverse Effects on Human Health.—Regulations issued under subparagraph (A) shall specify notification procedures for each violation, and each exceedance described in paragraph (1)(D), by a public water system that has the potential to have serious adverse effects on human health as a result of short-term exposure. Each notice of violation or exceedance provided under this subparagraph shall—

(i) be distributed as soon as practicable, but not later than 24 hours, after the public water system learns of the violation or exceedance;

(ii) provide a clear and readily understandable explanation of—

(I) the violation or exceedance;

(II) the potential adverse effects on human health;

(III) the steps that the public water system is taking to correct the violation or exceedance; and

(IV) the necessity of seeking alternative water supplies until the violation or exceedance is corrected;

(iii) be provided to the Administrator and the head of the State agency that has primary enforcement responsibility under section 1413, as applicable, as soon as practicable, but not later than 24 hours after the public water system learns of the violation or exceedance; and

(iv) as required by the State agency in general regulations of the State agency, or on a case-by-case basis after the consultation referred to in clause (iii), considering the health risks involved—

(I) be provided to appropriate media, including broadcast media;

(II) be prominently published in a newspaper of general circulation serving the area not later than 1 day after distribution of a notice pursuant to clause (i) or the date of publication of the next issue of the newspaper; or

(III) be provided by posting or door-to-door notification.

(D) Notice by the Administrator.—If the State with primary enforcement responsibility or the owner or operator of a public water system has not issued a notice under subparagraph (C) for an exceedance of the lead action level under section 141.80(c) of title 40, Code of Federal Regulations (or a prescribed level of lead that the Administrator establishes for public education or notification in a successor regulation promulgated pursuant to section 1412) that has the potential to have serious adverse effects on human health as a result of short-term exposure, not later than 24 hours after the Administrator is notified of the exceedance, the Administrator shall issue the required notice under that subparagraph.

(E) Written Notice.—

(i) In General.—Regulations issued under subparagraph (A) shall specify notification procedures for violations other than the violations covered by subparagraph (C). The procedures shall specify that a public water system shall provide written notice to each person served by the system by notice (I) in the first bill (if any) prepared after the date of occurrence of the violation, (II) in an annual report issued not later than 1 year after the date of occurrence of the violation, or (III) by mail or direct delivery as soon as practicable, but not later than 1 year after the date of occurrence of the violation.

(ii) Form and Manner of Notice.—The Administrator shall prescribe the form and manner of the notice to provide a clear and readily understandable explanation of the violation, any potential adverse health effects, and the steps that the system is taking to seek alternative water supplies, if any, until the violation is corrected.

(F) Unregulated Contaminants.—The Administrator may require the owner or operator of a public water system to give notice to the persons served by the system of the concentration levels of an unregulated contaminant required to be monitored under section 1445(a).

(3) Reports.—

(A) Annual Report By State.—

(i) In General.—Not later than January 1, 1998, and annually thereafter, each State that has primary enforcement responsibility under section

1413 shall prepare, make readily available to the public, and submit to the Administrator an annual report on violations of national primary drinking water regulations by public water systems in the State, including violations with respect to (I) maximum contaminant levels, (II) treatment requirements, (III) variances and exemptions, and (IV) monitoring requirements determined to be significant by the Administrator after consultation with the States.

(ii) Distribution.—The State shall publish and distribute summaries of the report and indicate where the full report is available for review.

(B) Annual Report By Administrator.—Not later than July 1, 1998, and annually thereafter, the Administrator shall prepare and make available to the public an annual report summarizing and evaluating reports submitted by States pursuant to subparagraph (A), notices submitted by public water systems serving Indian Tribes provided to the Administrator pursuant to subparagraph (C) or (E) of paragraph (2), and notices issued by the Administrator with respect to public water systems serving Indian Tribes under subparagraph (D) of that paragraph and making recommendations concerning the resources needed to improve compliance with this title. The report shall include information about public water system compliance on Indian reservations and about enforcement activities undertaken and financial assistance provided by the Administrator on Indian reservations, and shall make specific recommendations concerning the resources needed to improve compliance with this title on Indian reservations.

(4) Consumer Confidence Reports by Community Water Systems.—

(A) [Annual Reports] Reports to Consumers.—The Administrator, in consultation with public water systems, environmental groups, public interest groups, risk communication experts, and the States, and other interested parties, shall issue regulations within 24 months after the date of enactment of this paragraph to require each community water system to mail, or provide by electronic means, to each customer of the system at least once annually a report on the level of contaminants in the drinking water purveyed by that system (referred to in this paragraph as a "consumer confidence report"). Such regulations shall provide a brief and plainly worded definition of the terms "maximum contaminant level goal", "maximum contaminant level", "variances", and "exemptions" and brief statements in plain language regarding the health concerns that resulted in regulation of each regulated contaminant. The regulations shall also include a brief and plainly worded explanation regarding contaminants that may reasonably be expected to be present in drinking water, including bottled water. The regulations shall also provide for an Environmental

Protection Agency toll-free hotline that consumers can call for more information and explanation.

(B) Contents of Report.—The consumer confidence reports under this paragraph shall include, but not be limited to, each of the following:

(i) Information on the source of the water purveyed.

(ii) A brief and plainly worded definition of the terms “action level”, “maximum contaminant level goal”, “maximum contaminant level”, “variances”, and “exemptions” as provided in the regulations of the Administrator.

(iii) If any regulated contaminant is detected in the water purveyed by the public water system, a statement describing, as applicable—

(I) the maximum contaminant level goal;

(II) the maximum contaminant level;

(III) the level of the contaminant in the water system;

(IV) the action level for the contaminant; and

(V) for any contaminant for which there has been a violation of the maximum contaminant level during the year concerned, a brief statement in plain language regarding the health concerns that resulted in regulation of the contaminant, as provided by the Administrator in regulations under subparagraph (A).

(iv) Information on compliance with national primary drinking water regulations, as required by [the Administrator, and] the Administrator, including corrosion control efforts, and notice if the system is operating under a variance or exemption and the basis on which the variance or exemption was granted.

(v) Information on the levels of unregulated contaminants for which monitoring is required under section 1445(a)(2) (including levels of cryptosporidium and radon where States determine they may be found).

(vi) A statement that the presence of contaminants in drinking water does not necessarily indicate that the drinking water poses a health risk and that more information about contaminants and potential health effects can be obtained by calling the Environmental Protection Agency hotline.

(vii) Identification of, if any—

(I) exceedances described in paragraph (1)(D) for which corrective action has been required by the Administrator or the State (in the case of a State exercising primary enforcement responsibility for public water systems) during the monitoring period covered by the consumer confidence report; and

(II) violations that occurred during the monitoring period covered by the consumer confidence report.

A public water system may include such additional information as it deems appropriate for public education. The Administrator may, for not more than 3 regulated contaminants other than those referred to in clause (iii)(V), require a consumer confidence report under this paragraph to include the brief statement in plain language regarding the health concerns that resulted in regulation of the contaminant or contaminants concerned, as provided by the Administrator in regulations under subparagraph (A).

(C) COVERAGE.—The Governor of a State may determine not to apply the mailing requirement of subparagraph (A) to a community water system serving fewer than 10,000 persons. Any such system shall—

(i) inform, in the newspaper notice required by clause (iii) or by other means, its customers that the system will not be mailing the report as required by subparagraph (A);

(ii) make the consumer confidence report available upon request to the public; and

(iii) publish the report referred to in subparagraph (A) annually in one or more local newspapers serving the area in which customers of the system are located.

(D) Alternative to Publication.—For any community water system which, pursuant to subparagraph (C), is not required to meet the mailing requirement of subparagraph (A) and which serves 500 persons or fewer, the community water system may elect not to comply with clause (i) or (iii) of subparagraph (C). If the community water system so elects, the system shall, at a minimum—

(i) prepare an annual consumer confidence report pursuant to subparagraph (B); and

(ii) provide notice at least once per year to each of its customers by mail, by door-to-door delivery, by posting or by other means authorized by the regulations of the Administrator that the consumer confidence report is available upon request.

(E) Alternative Form and Content.—A State exercising primary enforcement responsibility may establish, by rule, after notice and public comment, alternative requirements with respect to the form and content of consumer confidence reports under this paragraph.

(F) Revisions.—

(i) Understandability And Frequency.—Not later than 24 months after the Drinking Water System Improvement Act of 2017, the Administrator, in consultation with the parties identified in subparagraph (A), shall issue revisions to the regulations issued under subparagraph (A)—

(I) to increase—(aa) the readability, clarity, and understandability of the information presented in consumer confidence reports; and (bb) the accuracy of information presented, and risk communication, in consumer confidence reports; and

(II) with respect to community water systems that serve 10,000 or more persons, to require each such community water system to provide, by mail, electronic means, or other methods described in clause (ii), a consumer confidence report to each customer of the system at least biannually.

(ii) Electronic Delivery.—Any revision of regulations pursuant to clause (i) shall allow delivery of consumer confidence reports by methods consistent with methods described in the memorandum "Safe Drinking Water Act–Consumer Confidence Report Rule Delivery Options" issued by the Environmental Protection Agency on January 3, 2013.

(5) Exceedance of Lead Level at Households.—

(A) Strategic Plan.—Not later than 180 days after the date of enactment of this paragraph, the Administrator shall, in collaboration with owners and operators of public water systems and States, establish a strategic plan for how the Administrator, a State with primary enforcement responsibility, and owners and operators of public water systems shall provide targeted outreach, education, technical assistance, and risk communication to populations affected by the concentration of lead in a public water system, including dissemination of information described in subparagraph (C).

(B) EPA Initiation of Notice.—

(i) Forwarding of Data by Employee of the Agency.—If the Agency develops, or receives from a source other than a State or a public water system, data that meets the requirements of section 1412(b)(3)(A)(ii) that indicates that the drinking water of a household served by a public water system contains a level of lead that exceeds the lead action level under section 141.80(c) of title 40, Code of Federal Regulations (or a prescribed level of lead that the Administrator establishes for public education or notification in a successor regulation promulgated pursuant to section 1412) (referred to in this paragraph as an "affected household"), the Administrator shall require an appropriate employee of the Agency to forward the data, and information on the sampling techniques used to obtain the data, to the owner or operator of the public water system and the State in which the affected household is located within a time period determined by the Administrator.

(ii) Dissemination of Information by Owner Or Operator.—The owner or operator of a public water system shall disseminate to affected households the information described in subparagraph (C) within a time period established by the Administrator, if the owner or operator—

(I) receives data and information under clause (i); and

(II) has not, since the date of the test that developed the data, notified the affected households—

(aa) with respect to the concentration of lead in the drinking water of the affected households; and

(bb) that the concentration of lead in the drinking water of the affected households exceeds the lead action level under section 141.80(c) of title 40, Code of Federal Regulations (or a prescribed level of lead that the Administrator establishes for public education or notification in a successor regulation promulgated pursuant to section 1412).

(iii) Consultation.—

(I) Deadline.—If the owner or operator of the public water system does not disseminate to the affected households the information described in subparagraph (C) as required under clause (ii) within the time period established by the Administrator, not later than 24 hours after the Administrator becomes aware of the failure by the owner or operator of the public water system to disseminate the information, the Administrator shall consult, within a period not to exceed 24 hours, with the applicable Governor to develop a plan, in accordance with the strategic plan, to disseminate the information to the affected households not later than 24 hours after the end of the consultation period.

(II) Delegation.—The Administrator may only delegate the duty to consult under subclause (I) to an employee of the Agency who, as of the date of the delegation, works in the Office of Water at the headquarters of the Agency.

(iv) Dissemination by Administrator.—The Administrator shall, as soon as practicable, disseminate to affected households the information described in subparagraph (C) if—

(I) the owner or operator of the public water system does not disseminate the information to the affected households within the time period determined by the Administrator, as required by clause (ii); and

(II)

(aa) the Administrator and the applicable Governor do not agree on a plan described in clause (iii)(I) during the consultation period under that clause; or

(bb) the applicable Governor does not disseminate the information within 24 hours after the end of the consultation period.

(C) Information Required.—The information described in this subparagraph includes—

(i) a clear explanation of the potential adverse effects on human health of drinking water that contains a concentration of lead that exceeds the lead action level under section 141.80(c) of title 40, Code of Federal Regulations (or a prescribed level of lead that the Administrator establishes for public education or notification in a successor regulation promulgated pursuant to section 1412);

(ii) the steps that the owner or operator of the public water system is taking to mitigate the concentration of lead; and

(iii) the necessity of seeking alternative water supplies until the date on which the concentration of lead is mitigated.

(6) Privacy.—Any notice to the public or an affected household under this subsection shall protect the privacy of individual customer information.

(d) Whenever, on the basis of information available to him, the Administrator finds that within a reasonable time after national secondary drinking water regulations have been promulgated, one or more public water systems in a State do not comply with such secondary regulations, and that such noncompliance appears to result from a failure of such State to take reasonable action to assure that public water systems throughout such State meet such secondary regulations, he shall so notify the State.

(e) Nothing in this title shall diminish any authority of a State or political subdivision to adopt or enforce any law or regulation respecting drinking water regulations or public water systems, but no such law or regulation shall relieve any person of any requirement otherwise applicable under this title.

(f) If the Administrator makes a finding of noncompliance (de- scribed in subparagraph (A) or (B) of subsection (a)(1)) with respect to a public water system in a State which has primary enforcement responsibility, the Administrator may, for the purpose of assisting that State in carrying out such responsibility and upon the petition of such State or public water system or persons served by such system, hold, after appropriate notice, public hearings for the purpose of gathering information from technical or other experts, Federal, State, or other public officials, representatives of such public water system, persons served by such system, and other interested persons on—

(1) the ways in which such system can within the earliest feasible time be brought into compliance with the regulation or requirement with respect to which such finding was made, and

(2) the means for the maximum feasible protection of the public health during any period in which such system is not in compliance with a national primary drinking water regulation or requirement applicable to a variance or exemption.

On the basis of such hearings the Administrator shall issue recommendations which shall be sent to such State and public water system and shall be made available to the public and communications media.

(g)

(1) In any case in which the Administrator is authorized to bring a civil action under this section or under section 1445 with respect to any applicable requirement, the Administrator also may issue an order to require compliance with such applicable requirement.

(2) An order issued under this subsection shall not take effect, in the case of a State having primary enforcement responsibility for public water systems in that State, until after the Administrator has provided the State with an opportunity to confer with the Administrator regarding the order. A copy of any order issued under this subsection shall be sent to the appropriate State agency of the State involved if the State has primary enforcement responsibility for public water systems in that State. Any order issued under this subsection shall state with reasonable specificity the nature of the violation. In any case in which an order under this subsection is issued to a corporation, a copy of such order shall be issued to appropriate corporate officers.

(3)

(A) Any person who violates, or fails or refuses to comply with, an order under this subsection shall be liable to the United States for a civil penalty of not more than $25,000 per day of violation.

(B) In a case in which a civil penalty sought by the Administrator under this paragraph does not exceed $5,000, the penalty shall be assessed by the Administrator after notice and opportunity for a public hearing (unless the person against whom the penalty is assessed requests a hearing on the record in accordance with section 554 of title 5, United States Code). In a case in which a civil penalty sought by the Administrator under this paragraph exceeds $5,000, but does not exceed $25,000, the penalty shall be assessed by the Administrator after notice and opportunity for a hearing on the record in accordance with section 554 of title 5, United States Code.

(C) Whenever any civil penalty sought by the Administrator under this subsection for a violation of an applicable requirement exceeds $25,000, the penalty shall be assessed by a civil action brought by the Administrator in the appropriate United States district court (as determined under the provisions of title 28 of the United States Code).

(D) If any person fails to pay an assessment of a civil penalty after it has become a final and unappealable order, or after the appropriate court of appeals has entered final judgment in favor of the Administrator, the Attorney General shall recover the amount for which such person is liable in any appropriate district court of the United States. In any such action, the validity and appropriateness of the final order imposing the civil penalty shall not be subject to review.

(h) Consolidation Incentive.—

(1) In General.—An owner or operator of a public water system may submit to the State in which the system is located (if the State has primary enforcement responsibility under section 1413) or to the Administrator (if the State does not have primary enforcement responsibility) a plan (including specific measures and schedules) for—

(A) the physical consolidation of the system with 1 or more other systems;

(B) the consolidation of significant management and administrative functions of the system with 1 or more other systems; [or]

(C) the transfer of ownership of the system that may reasonably be expected to improve drinking water quality[.]; or

(D) entering into a contractual agreement for significant management or administrative functions of the system to correct violations identified in the plan.

(2) Consequences of Approval.—If the State or the Administrator approves a plan pursuant to paragraph (1), no enforcement action shall be taken pursuant to this part with respect to a specific violation identified in the approved plan prior to the date that is the earlier of the date on which consolidation is completed according to the plan or the date that is 2 years after the plan is approved.

(3) Authority for Mandatory Assessment and Mandatory Consolidation.—

(A) Mandatory Assessment.—A State with primary enforcement responsibility or the Administrator (if the State does not have primary enforcement responsibility) may require the owner or operator of a public water system to assess options for consolidation, or transfer of ownership of the system, as described in paragraph (1), if—

(i) the public water system—

(I) has repeatedly violated one or more national primary drinking water regulations and such repeated violations are likely to adversely affect human health; and

(II)

(aa) is unable or unwilling to take feasible and affordable actions, as identified by the State with primary enforcement responsibility or the Administrator (if the State does not have primary enforcement responsibility), that will result in the public water system complying with the national primary drinking water regulations described in subclause

(I) including accessing technical assistance and financial assistance through the State loan fund pursuant to section 1452; or

(bb) has already undertaken actions described in item (aa) without achieving compliance;

(ii) such consolidation or transfer is feasible; and

(iii) such consolidation or transfer could result in greater compliance with national primary drinking water regulations.

(B) Mandatory Consolidation.—After review of an assessment under subparagraph (A), a State with primary enforcement responsibility or the Administrator (if the State does not have primary enforcement responsibility) may require the owner or operator of a public water system that completed such assessment to submit a plan for consolidation, or transfer of ownership of the system, under paragraph (1), and complete the actions required under such plan if—

(i) the owner or operator of the public water system—

(I) has not taken steps to complete consolidation;

(II) has not transferred ownership of the system; or

(III) was unable to achieve compliance after taking the actions described in clause (i)(II)(aa) of subparagraph (A);

(ii) since completing such assessment, the public water system has violated one or more national primary drinking water regulations and such violations are likely to adversely affect human health; and

(iii) such consolidation or transfer is feasible.

(4) Financial Assistance.—Notwithstanding section 1452(a)(3), a public water system undertaking consolidation or transfer of ownership or alternative actions to achieve compliance pursuant to this subsection may receive assistance under section 1452 to carry out such consolidation, transfer, or alternative actions.

(5) Protection of Nonresponsible System.—

(A) Identification of Liabilities.—

(i) In General.—An owner or operator of a public water system submitting a plan pursuant to paragraph (3) shall identify as part of such plan—

(I) any potential liability for damages arising from each specific violation identified in the plan of which the owner or operator is aware; and

(II) any funds or other assets that are available to satisfy such liability, as of the date of submission of such plan, to the public water system that committed such violation.

(ii) Inclusion.—In carrying out clause (i), the owner or operator shall take reasonable steps to ensure that all potential liabilities for damages arising from each specific violation identified in the plan submitted pursuant to paragraph (3) are identified.

(B) Reservation of Funds.—A public water system that has completed the actions required under a plan submitted and approved pursuant to paragraph (3) shall not be liable under this title for a violation of this title identified in the plan, except to the extent to which funds or other assets are identified pursuant to subparagraph (A)(i)(II) as available to satisfy such liability.

(6) Regulations.—Not later than 2 years after the date of enactment of the Drinking Water System Improvement Act of 2017, the Administrator shall promulgate regulations to implement paragraphs (3), (4), and (5).

(i) Definition of Applicable Requirement.—In this section, the term "applicable requirement" means—

(1) a requirement of section 1412, 1414, 1415, 1416, 1417, 1433, 1441, or 1445;

(2) a regulation promulgated pursuant to a section referred to in paragraph (1);

(3) a schedule or requirement imposed pursuant to a section referred to in paragraph (1); and

(4) a requirement of, or permit issued under, an applicable State program for which the Administrator has made a determination that the requirements of section 1413 have been satisfied, or an applicable State program approved pursuant to this part.

(j) Improved Accuracy and Availability of Compliance Monitoring Data.—

(1) Strategic Plan.—Not later than 1 year after the date of enactment of this subsection, the Administrator, in coordination with States, public water systems, and other interested stakeholders, shall develop and provide to Congress a strategic plan for improving the accuracy and availability of monitoring data collected to demonstrate compliance with national primary drinking water regulations and submitted—

(A) by public water systems to States; or

(B) by States to the Administrator.

(2) Evaluation.—In developing the strategic plan under paragraph (1), the Administrator shall evaluate any challenges faced—

(A) in ensuring the accuracy and integrity of submitted data described in paragraph (1);

(B) by States and public water systems in implementing an electronic system for submitting such data, including the technical and economic feasibility of implementing such a system; and

(C) by users of such electronic systems in being able to access such data.

(3) Findings and Recommendations.—The Administrator shall include in the strategic plan provided to Congress under paragraph (1)—

(A) a summary of the findings of the evaluation under paragraph (2); and

(B) recommendations on practicable, cost-effective methods and means that can be employed to improve the accuracy and availability of submitted data described in paragraph (1).

(4) CONSULTATION.—In developing the strategic plan under paragraph (1), the Administrator may, as appropriate, consult with States or other Federal agencies that have experience using practicable methods and means to improve the accuracy and availability of submitted data described in such paragraph.

Capacity Development

Sec. 1420.

(a) State Authority for New Systems.—A State shall receive only 80 percent of the allotment that the State is otherwise entitled to receive under section 1452 (relating to State loan funds) unless the State has obtained the legal authority or other means to ensure that all new community water systems and new nontransient, noncommunity water systems commencing operation after October 1, 1999, demonstrate technical, managerial, and financial capacity with respect to each national primary drinking water regulation in effect, or likely to be in effect, on the date of commencement of operations.

(b) Systems in Significant Noncompliance.—

(1) List.—Beginning not later than 1 year after the date of enactment of this section, each State shall prepare, periodically update, and submit to the Administrator a list of community water systems and nontransient, noncommunity water systems that have a history of significant noncompliance with this title (as defined in guidelines issued prior to the date of enactment of this section or any revisions of the guidelines that have been made in

consultation with the States) and, to the extent practicable, the reasons for noncompliance.

(2) Report.—Not later than 5 years after the date of enactment of this section and as part of the capacity development strategy of the State, each State shall report to the Administrator on the success of enforcement mechanisms and initial capacity development efforts in assisting the public water systems listed under paragraph (1) to improve technical, managerial, and financial capacity.

(3) Withholding.—The list and report under this subsection shall be considered part of the capacity development strategy of the State required under subsection (c) of this section for purposes of the withholding requirements of section 1452(a)(1)(G)(i) (relating to State loan funds).

(c) Capacity Development Strategy.—

(1) In General.—Beginning 4 years after the date of enactment of this section, a State shall receive only—

(A) 90 percent in fiscal year 2001;

(B) 85 percent in fiscal year 2002; and

(C) 80 percent in each subsequent fiscal year, of the allotment that the State is otherwise entitled to receive under section 1452 (relating to State loan funds), unless the State is developing and implementing a strategy to assist public water systems in acquiring and maintaining technical, managerial, and financial capacity.

(2) Content.—In preparing the capacity development strategy, the State shall consider, solicit public comment on, and include as appropriate—

(A) the methods or criteria that the State will use to identify and prioritize the public water systems most in need of improving technical, managerial, and financial capacity;

(B) a description of the institutional, regulatory, financial, tax, or legal factors at the Federal, State, or local level that encourage or impair capacity development;

(C) a description of how the State will use the authorities and resources of this title or other means to—

(i) assist public water systems in complying with national primary drinking water regulations;

(ii) encourage the development of partnerships between public water systems to enhance the technical, managerial, and financial capacity of the systems; and

(iii) assist public water systems in the training and certification of operators;

(D) a description of how the State will establish a baseline and measure improvements in capacity with respect to national primary drinking water regulations and State drinking water law[; and];

(E) an identification of the persons that have an interest in and are involved in the development and implementation of the capacity development strategy (including all appropriate agencies of Federal, State, and local governments, private and nonprofit public water systems, and public water system customers)[.]; and

(F) a description of how the State will, as appropriate—

(i) encourage development by public water systems of asset management plans that include best practices for asset management; and

(ii) assist, including through the provision of technical assistance, public water systems in training operators or other relevant and appropriate persons in implementing such asset management plans.

(3) Report.—Not later than 2 years after the date on which a State first adopts a capacity development strategy under this subsection, and every 3 years thereafter, the head of the State agency that has primary responsibility to carry out this title in the State shall submit to the Governor a report that shall also be available to the public on the efficacy of the strategy and progress made toward improving the technical, managerial, and financial capacity of public water systems in the State, including efforts of the State to encourage development by public water systems of asset management plans and to assist public water systems in training relevant and appropriate persons in implementing such asset management plans.

(4) Review.—The decisions of the State under this section regarding any particular public water system are not subject to review by the Administrator and may not serve as the basis for withholding funds under section 1452.

(d) Federal Assistance.—

(1) In General.—The Administrator shall support the States in developing capacity development strategies.

(2) Informational Assistance.—

(A) IN GENERAL.—Not later than 180 days after the date of enactment of this section, the Administrator shall—

(i) conduct a review of State capacity development efforts in existence on the date of enactment of this section and publish information to assist States and public water systems in capacity development efforts; and

(ii) initiate a partnership with States, public water systems, and the public to develop information for States on recommended operator certification requirements.

(B) Publication of Information.—The Administrator shall publish the information developed through the partnership under subparagraph (A)(ii) not later than 18 months after the date of enactment of this section.

(3) Promulgation of Drinking Water Regulations.—In promulgating a national primary drinking water regulation, the Administrator shall include an analysis of the likely effect of compliance with the regulation on the technical, financial, and managerial capacity of public water systems.

(4) Guidance for New Systems.—Not later than 2 years after the date of enactment of this section, the Administrator shall publish guidance developed in consultation with the States describing legal authorities and other means to ensure that all new community water systems and new nontransient, noncommunity water systems demonstrate technical, managerial, and financial capacity with respect to national primary drinking water regulations.

(5) Information on Asset Management Practices.—Not later than 5 years after the date of enactment of this paragraph, and not less often than every 5 years thereafter, the Administrator shall review and, if appropriate, update educational materials, including handbooks, training materials, and technical information, made available by the Administrator to owners, managers, and operators of public water systems, local officials, technical assistance providers (including nonprofit water associations), and State personnel concerning best practices for asset management strategies that may be used by public water systems.

(e) Variances and Exemptions.—Based on information obtained under subsection (c)(3), the Administrator shall, as appropriate, modify regulations concerning variances and exemptions for small public water systems to ensure flexibility in the use of the variances and exemptions. Nothing in this subsection shall be interpreted, construed, or applied to affect or alter the requirements of section 1415 or 1416.

(f) Small Public Water Systems Technology Assistance Centers.—

(1) Grant Program.—The Administrator is authorized to make grants to institutions of higher learning to establish and operate small public water system technology assistance centers in the United States.

(2) Responsibilities of the Centers.—The responsibilities of the small public water system technology assistance centers established under this subsection shall include the conduct of training and technical assistance relating to the information, performance, and technical needs of small public water systems or public water systems that serve Indian Tribes.

(3) Applications.—Any institution of higher learning interested in receiving a grant under this subsection shall submit to the Administrator an application in

such form and containing such information as the Administrator may require by regulation.

(4) Selection Criteria.—The Administrator shall select recipients of grants under this subsection on the basis of the following criteria:

(A) The small public water system technology assistance center shall be located in a State that is representative of the needs of the region in which the State is located for addressing the drinking water needs of small and rural communities or Indian Tribes.

(B) The grant recipient shall be located in a region that has experienced problems, or may reasonably be foreseen to experience problems, with small and rural public water systems.

(C) The grant recipient shall have access to expertise in small public water system technology management.

(D) The grant recipient shall have the capability to disseminate the results of small public water system technology and training programs.

(E) The projects that the grant recipient proposes to carry out under the grant are necessary and appropriate.

(F) The grant recipient has regional support beyond the host institution.

(5) Consortia of States.—At least 2 of the grants under this subsection shall be made to consortia of States with low population densities.

(6) Authorization of Appropriations.—There are authorized to be appropriated to make grants under this subsection $2,000,000 for each of the fiscal years 1997 through 1999, and $5,000,000 for each of the fiscal years 2000 through 2003.

(g) Environmental Finance Centers.—

(1) In General.—The Administrator shall provide initial funding for one or more university-based environmental finance centers for activities that provide technical assistance to State and local officials in developing the capacity of public water systems. Any such funds shall be used only for activities that are directly related to this title.

(2) National Capacity Development Clearinghouse.—The Administrator shall establish a national public water system capacity development clearinghouse to receive and disseminate information with respect to developing, improving, and maintaining financial and managerial capacity at public water systems. The Administrator shall ensure that the clearinghouse does not duplicate other federally supported clearinghouse activities.

(3) Capacity Development Techniques.—The Administrator may request an environmental finance center funded under paragraph (1) to develop and test managerial, financial, and institutional techniques for capacity development. The techniques may include capacity assessment methodologies, manual and

computer based public water system rate models and capital planning models, public water system consolidation procedures, and regionalization models.

(4) Authorization of Appropriations.—There are authorized to be appropriated to carry out this subsection $1,500,000 for each of the fiscal years 1997 through 2003.

(5) Limitation.—No portion of any funds made available under this subsection may be used for lobbying expenses.

Part D—Emergency Powers

[Sec. 1433.] Terrorist and Other Intentional Acts.

[(a) Vulnerability Assessments.—

(1) Each community water system serving a population of greater than 3,300 persons shall conduct an assessment of the vulnerability of its system to a terrorist attack or other intentional acts intended to substantially disrupt the ability of the system to provide a safe and reliable supply of drinking water. The vulnerability assessment shall include, but not be limited to, a review of pipes and constructed conveyances, physical barriers, water collection, pretreatment, treatment, storage and distribution facilities, electronic, computer or other automated systems which are utilized by the public water system, the use, storage, or handling of various chemicals, and the operation and maintenance of such system. The Administrator, not later than August 1, 2002, after consultation with appropriate departments and agencies of the Federal Government and with State and local governments, shall provide baseline information to community water systems required to conduct vulnerability assessments regarding which kinds of terrorist attacks or other intentional acts are the probable threats to—

[(A) substantially disrupt the ability of the system to provide a safe and reliable supply of drinking water; or

[(B) otherwise present significant public health concerns.

[(2) Each community water system referred to in paragraph (1) shall certify to the Administrator that the system has conducted an assessment complying with paragraph (1) and shall submit to the Administrator a written copy of the assessment. Such certification and submission shall be made prior to:

[(A) March 31, 2003, in the case of systems serving a population of 100,000 or more.

[(B) December 31, 2003, in the case of systems serving a population of 50,000 or more but less than 100,000.

[(C) June 30, 2004, in the case of systems serving a population greater than 3,300 but less than 50,000.

[(3) Except for information contained in a certification under this subsection identifying the system submitting the certification and the date of the certification, all information provided to the Administrator under this subsection and all information derived therefrom shall be exempt from disclosure under section 552 of title 5 of the United States Code.

[(4) No community water system shall be required under State or local law to provide an assessment described in this section to any State, regional, or local governmental entity solely by reason of the requirement set forth in paragraph (2) that the system submit such assessment to the Administrator.

[(5) Not later than November 30, 2002, the Administrator, in consultation with appropriate Federal law enforcement and intelligence officials, shall develop such protocols as may be necessary to protect the copies of the assessments required to be submitted under this subsection (and the information contained therein) from unauthorized disclosure. Such protocols shall ensure that—

[(A) each copy of such assessment, and all information contained in or derived from the assessment, is kept in a secure location;

[(B) only individuals designated by the Administrator may have access to the copies of the assessments; and

[(C) no copy of an assessment, or part of an assessment, or information contained in or derived from an assessment shall be available to anyone other than an individual designated by the Administrator.

At the earliest possible time prior to November 30, 2002, the Administrator shall complete the development of such protocols for the purpose of having them in place prior to receiving any vulnerability assessments from community water systems under this subsection.

[(6)

(A) Except as provided in subparagraph (B), any individual referred to in paragraph (5)(B) who acquires the assessment submitted under paragraph (2), or any reproduction of such assessment, or any information derived from such assessment, and who knowingly or recklessly reveals such assessment, reproduction, or information other than—

[(i) to an individual designated by the Administrator under paragraph (5),

[(ii) for purposes of section 1445 or for actions under section 1431, or

[(iii) for use in any administrative or judicial proceeding to impose a penalty for failure to comply with this section, shall upon conviction be imprisoned for not more than one year or fined in accordance with the provisions of chapter 227 of title 18, United States Code, applicable to class A misdemeanors, or both, and shall be removed from Federal office or employment.

[(B) Notwithstanding subparagraph (A), an individual referred to in paragraph (5)(B) who is an officer or employee of the United States may discuss the contents of a vulnerability assessment submitted under this section with a State or local official.

[(7) Nothing in this section authorizes any person to withhold any information from Congress or from any committee or subcommittee of Congress.

[(b) Emergency Response Plan.—Each community water system serving a population greater than 3,300 shall prepare or revise, where necessary, an emergency response plan that incorporates the results of vulnerability assessments that have been completed. Each such community water system shall certify to the Administrator, as soon as reasonably possible after the enactment of this section, but not later than 6 months after the completion of the vulnerability assessment under subsection (a), that the system has completed such plan. The emergency response plan shall include, but not be limited to, plans, procedures, and identification of equipment that can be implemented or utilized in the event of a terrorist or other intentional attack on the public water system. The emergency response plan shall also include actions, procedures, and identification of equipment which can obviate or significantly lessen the impact of terrorist attacks or other intentional actions on the public health and the safety and supply of drinking water provided to communities and individuals. Community water systems shall, to the extent possible, coordinate with existing Local Emergency Planning Committees established under the Emergency Planning and Community Right-to-Know Act (42 U.S.C. 11001 et seq.) when preparing or revising an emergency response plan under this subsection.

[(c) Record Maintenance.—Each community water system shall maintain a copy of the emergency response plan completed pursuant to subsection (b) for 5 years after such plan has been certified to the Administrator under this section.

[(d) Guidance to Small Public Water Systems.—The Administrator shall provide guidance to community water systems serving a population of less than 3,300 persons on how to conduct vulnerability assessments, prepare emergency response plans, and address threats from terrorist attacks or other intentional actions designed to disrupt the provision of safe drinking water or significantly affect the public health or significantly affect the safety or supply of drinking water provided to communities and individuals.

[(e) Funding.—

(1) There are authorized to be appropriated to carry out this section not more than $160,000,000 for the fiscal year 2002 and such sums as may be necessary for the fiscal years 2003 through 2005.

[(2) The Administrator, in coordination with State and local governments, may use funds made available under paragraph (1) to provide financial assistance to community water systems for purposes of compliance with the requirements of subsections (a) and (b) and to community water systems for expenses and contracts designed to address basic security enhancements of critical importance and significant threats to public health and the supply of drinking water as determined by a vulnerability assessment conducted under subsection (a). Such basic security enhancements may include, but shall not be limited to the following:

[(A) the purchase and installation of equipment for detection of intruders;

[(B) the purchase and installation of fencing, gating, lighting, or security cameras;

[(C) the tamper-proofing of manhole covers, fire hydrants, and valve boxes;

[D. the rekeying of doors and locks;

[(E) improvements to electronic, computer, or other automated systems and remote security systems;

[(F) participation in training programs, and the purchase of training manuals and guidance materials, relating to security against terrorist attacks;

[(G) improvements in the use, storage, or handling of various chemicals; and

[(H) security screening of employees or contractor support services.

Funding under this subsection for basic security enhancements shall not include expenditures for personnel costs, or monitoring, operation, or maintenance of facilities, equipment, or systems.

[(3) The Administrator may use not more than $5,000,000 from the funds made available under paragraph (1) to make grants to community water systems to assist in responding to and alleviating any vulnerability to a terrorist attack or other intentional acts intended to substantially disrupt the ability of the system to provide a safe and reliable supply of drinking water (including sources of water for such systems) which the Administrator determines to present an immediate and urgent security need.

[(4) The Administrator may use not more than $5,000,000 from the funds made available under paragraph (1) to make grants to community water systems serving a population of less than 3,300 persons for activities and projects undertaken in accordance with the guidance provided to such systems under subsection (d).]

Sec. 1433. Community Water System Risk and Resilience.

(a) Risk and Resilience Assessments.—

(1) In General.—Each community water system serving a population of greater than 3,300 persons shall conduct an assessment of the risks to, and resilience of, its system. Such an assessment—

(A) shall include an assessment of—

(i) the risk to the system from malevolent acts and natural hazards;

(ii) the resilience of the pipes and constructed conveyances, physical barriers, source water, water collection and intake, pretreatment, treatment, storage and distribution facilities, electronic, computer, or other automated systems (including the security of such systems) which are utilized by the system;

(iii) the monitoring practices of the system; (iv) the financial infrastructure of the system;

(v) the use, storage, or handling of various chemicals by the system; and

(vi) the operation and maintenance of the system; and

(B) may include an evaluation of capital and operational needs for risk and resilience management for the system.

(2) Baseline Information.—The Administrator, not later than August 1, 2019, after consultation with appropriate departments and agencies of the Federal Government and with State and local governments, shall provide baseline information on malevolent acts of relevance to community water systems, which shall include consideration of acts that may—

(A) substantially disrupt the ability of the system to provide a safe and reliable supply of drinking water; or

(B) otherwise present significant public health or economic concerns to the community served by the system.

(3) CERTIFICATION.—

(A) Certification.—Each community water system described in paragraph (1) shall submit to the Administrator a certification that the system has conducted an assessment complying with paragraph (1). Such certification shall be made prior to—

(i) March 31, 2020, in the case of systems serving a population of 100,000 or more;

(ii) December 31, 2020, in the case of systems serving a population of 50,000 or more but less than 100,000; and

(iii) June 30, 2021, in the case of systems serving a population greater than 3,300 but less than 50,000.

(B) Review and Revision.—Each community water system described in paragraph (1) shall review the assessment of such system conducted under such paragraph at least once every 5 years after the applicable deadline for submission of its certification under subparagraph (A) to determine whether such assessment should be revised. Upon completion of such a review, the community water system shall submit to the Administrator a certification that the system has reviewed its assessment and, if applicable, revised such assessment.

(4) Contents of Certifications.—A certification required under paragraph (3) shall contain only—

(A) information that identifies the community water system submitting the certification;

(B) the date of the certification; and

(C) a statement that the community water system has conducted, reviewed, or revised the assessment, as applicable.

(5) Provision to Other Entities.—No community water system shall be required under State or local law to provide an assessment described in this section (or revision thereof) to any State, regional, or local governmental entity solely by reason of the requirement set forth in paragraph (3) that the system submit a certification to the Administrator.

(b) Emergency Response Plan.—Each community water system serving a population greater than 3,300 shall prepare or revise, where necessary, an emergency response plan that incorporates findings of the assessment conducted under subsection (a) for such system (and any revisions thereto). Each community water system shall certify to the Administrator, as soon as reasonably possible after the date of enactment of the Drinking Water System Improvement Act of 2017, but not later than 6 months after completion of the assessment under subsection (a), that the system has completed such plan. The emergency response plan shall include—

(1) strategies and resources to improve the resilience of the system, including the physical security and cybersecurity of the system;

(2) plans and procedures that can be implemented, and identification of equipment that can be utilized, in the event of a malevolent act or natural hazard that threatens the ability of the community water system to deliver safe drinking water;

(3) actions, procedures, and equipment which can obviate or significantly lessen the impact of a malevolent act or natural hazard on the public health and the safety and supply of drinking water provided to communities and individuals, including the development of alternative source water options, relocation of water intakes, and construction of flood protection barriers; and

(4) strategies that can be used to aid in the detection of malevolent acts or natural hazards that threaten the security or resilience of the system.

(c) Coordination.—Community water systems shall, to the extent possible, coordinate with existing local emergency planning committees established pursuant to the

Emergency Planning and Community Right-To-Know Act of 1986 (42 U.S.C. 11001 et seq.) when preparing or revising an assessment or emergency response plan under this section.

(d) Record Maintenance.—Each community water system shall maintain a copy of the assessment conducted under subsection (a) and the emergency response plan prepared under subsection (b) (including any revised assessment or plan) for 5 years after the date on which a certification of such assessment or plan is submitted to the Administrator under this section.

(e) Guidance to Small Public Water Systems.—The Administrator shall provide guidance and technical assistance to community water systems serving a population of less than 3,300 persons on how to conduct resilience assessments, prepare emergency response plans, and address threats from malevolent acts and natural hazards that threaten to disrupt the provision of safe drinking water or significantly affect the public health or significantly affect the safety or supply of drinking water provided to communities and individuals.

(f) Alternative Preparedness and Operational Resilience Programs.—

(1) Satisfaction of Requirement.—A community water system that is required to comply with the requirements of subsections (a) and (b) may satisfy such requirements by—

(A) using and complying with technical standards that the Administrator has recognized under paragraph (2); and

(B) submitting to the Administrator a certification that the community water system is complying with subparagraph (A).

(2) Authority to Recognize.—Consistent with section 12(d) of the National Technology Transfer and Advancement Act of 1995, the Administrator shall recognize technical standards that are developed or adopted by third-party organizations or voluntary consensus standards bodies that carry out the objecment under subsection (a), that the system has completed such plan. fying the requirements under subsection (a) or (b).

(g) Technical Assistance and Grants.—

(1) In General.—The Administrator shall establish and implement a program, to be known as the Drinking Water Infrastructure Risk and Resilience Program, under which the Administrator may award grants in each of fiscal years 2018 through 2022 to owners or operators of community water systems for the purpose of increasing the resilience of such community water systems.

(2) Use of Funds.—As a condition on receipt of a grant under this section, an owner or operator of a community water system shall agree to use the grant funds exclusively to assist in the planning, design, construction, or implementation of a program or project consistent with an emergency response plan prepared pursuant to subsection (b), which may include—

(A) the purchase and installation of equipment for detection of drinking water contaminants or malevolent acts;

(B) the purchase and installation of fencing, gating, lighting, or security cameras;

(C) the tamper-proofing of manhole covers, fire hydrants, and valve boxes;

(D) the purchase and installation of improved treatment technologies and equipment to improve the resilience of the system;

(E) improvements to electronic, computer, financial, or other automated systems and remote systems;

(F) participation in training programs, and the purchase of training manuals and guidance materials, relating to security and resilience;

(G) improvements in the use, storage, or handling of chemicals by the community water system;

(H) security screening of employees or contractor support services;

(I) equipment necessary to support emergency power or water supply, including standby and mobile sources; and

(J) the development of alternative source water options, relocation of water intakes, and construction of flood protection barriers.

(3) EXCLUSIONS.—A grant under this subsection may not be used for personnel costs, or for monitoring, operation, or maintenance of facilities, equipment, or systems.

(4) TECHNICAL ASSISTANCE.—For each fiscal year, the Administrator may use not more than $5,000,000 from the funds made available to carry out this subsection to provide technical assistance to community water systems to assist in responding to and alleviating a vulnerability that would substantially disrupt the ability of the system to provide a safe and reliable supply of drinking water (including sources of water for such systems) which the Administrator determines to present an immediate and urgent need.

(5) GRANTS FOR SMALL SYSTEMS.—For each fiscal year, the Administrator may use not more than $10,000,000 from the funds made available to carry out this subsection to make grants to community water systems serving a population of less than 3,300 persons, or nonprofit organizations receiving assistance under section 1442(e), for activities and projects undertaken in accordance with the guidance provided to such systems under subsection (e) of this section.

(6) Authorization of Appropriations.—To carry out this subsection, there are authorized to be appropriated $35,000,000 for each of fiscal years 2018 through 2022.

(h) Definitions.—In this section—

(1) the term "resilience" means the ability of a community water system or an asset of a community water system to adapt to or withstand the effects of a malevolent act or natural hazard without interruption to the asset's or system's function, or if the function is interrupted, to rapidly return to a normal operating condition; and

(2) the term "natural hazard" means a natural event that threatens the functioning of a community water system, including an earthquake, tornado, flood, hurricane, wildfire, and hydrologic changes.

Part E—General Provisions

Grants for State Programs

Sec. 1443.

(a)

(1) From allotments made pursuant to paragraph (4), the Administrator may make grants to States to carry out public water system supervision programs.

(2) No grant may be made under paragraph (1) unless an application therefor has been submitted to the Administrator in such form and manner as he may require. The Administrator may not approve an application of a State for its first grant under paragraph (1) unless he determines that the State—

(A) has established or will establish within one year from the date of such grant a public water system supervision program, and

(B) will, within that one year, assume primary enforcement responsibility for public water systems within the State.

No grant may be made to a State under paragraph (1) for any period beginning more than one year after the date of the State's first grant unless the State has assumed and maintains primary enforcement responsibility for public water systems within the State. The prohibitions contained in the preceding two sentences shall not apply to such grants when made to Indian Tribes.

(3) A grant under paragraph (1) shall be made to cover not more than 75 per centum of the grant recipient's costs (as determined under regulations of the Administrator) in carrying out, during the one-year period beginning on the date the grant is made, a public water system supervision program.

(4) In each fiscal year the Administrator shall, in accordance with regulations, allot the sums appropriated for such year under paragraph (5) among the States on the basis of population, geographical area, number of public water systems, and other relevant factors. No State shall receive less than 1 per centum of the

annual appropriation for grants under paragraph (1): *Provided,* That the Administrator may, by regulation, reduce such percentage in accordance with the criteria specified in this paragraph: *And provided further,* That such percentage shall not apply to grants allotted to Guam, American Samoa, or the Virgin Islands.

(5) The prohibition contained in the last sentence of paragraph (2) may be waived by the Administrator with respect to a grant to a State through fiscal year 1979 but such prohibition may only be waived if, in the judgment of the Administrator—

(A) the State is making a diligent effort to assume and maintain primary enforcement responsibility for public water systems within the State;

(B) the State has made significant progress toward assuming and maintaining such primary enforcement responsibility; and

(C) there is reason to believe the State will assume such primary enforcement responsibility by October 1, 1979.

The amount of any grant awarded for the fiscal years 1978 and 1979 pursuant to a waiver under this paragraph may not exceed 75 per centum of the allotment which the State would have received for such fiscal year if it had assumed and maintained such primary enforcement responsibility. The remaining 25 per centum of the amount allotted to such State for such fiscal year shall be retained by the Administrator, and the Administrator may award such amount to such State at such time as the State assumes such responsibility before the beginning of fiscal year 1980. At the beginning of each fiscal years 1979 and 1980 the amounts retained by the Administrator for any preceding fiscal year and not awarded by the beginning of fiscal year 1979 or 1980 to the States to which such amounts were originally allotted may be removed from the original allotment and reallotted for fiscal year 1979 or 1980 (as the case may be) to States which have assumed primary enforcement responsibility by the beginning of such fiscal year.

(6) The Administrator shall notify the State of the approval or disapproval of any application for a grant under this section—

(A) within ninety days after receipt of such application, or

(B) not later than the first day of the fiscal year for which the grant application is made, whichever is later.

(7) Authorization.—For the purpose of making grants under paragraph (1), there are authorized to be appropriated [$100,000,000 for each of fiscal years 1997 through 2003] *$150,000,000 for each of fiscal years 2018 through 2022.*

(8) Reservation of Funds by the Administrator.—If the Administrator assumes the primary enforcement responsibility of a State public water system supervision program, the Administrator may reserve from funds made available pursuant

to this subsection an amount equal to the amount that would otherwise have been provided to the State pursuant to this subsection. The Administrator shall use the funds reserved pursuant to this paragraph to ensure the full and effective administration of a public water system supervision program in the State.

(9) State Loan Funds.—

(A) Reservation of Funds.—For any fiscal year for which the amount made available to the Administrator by appropriations to carry out this subsection is less than the amount that the Administrator determines is necessary to supplement funds made available pursuant to paragraph (8) to ensure the full and effective administration of a public water system supervision program in a State, the Administrator may reserve from the funds made available to the State under section 1452 (relating to State loan funds) an amount that is equal to the amount of the shortfall. This paragraph shall not apply to any State not exercising primary enforcement responsibility for public water systems as of the date of enactment of the Safe Drinking Water Act Amendments of 1996.

(B) Duty of Administrator.—If the Administrator reserves funds from the allocation of a State under subparagraph (A), the Administrator shall carry out in the State each of the activities that would be required of the State if the State had primary enforcement authority under section 1413.

(b)

(1) From allotments made pursuant to paragraph (4), the Administrator may make grants to States to carry out underground water source protection programs.

(2) No grant may be made under paragraph (1) unless an application therefor has been submitted to the Administrator in such form and manner as he may require. No grant may be made to any State under paragraph (1) unless the State has assumed primary enforcement responsibility within two years after the date the Administrator promulgates regulations for State underground injection control programs under section 1421. The prohibition contained in the preceding sentence shall not apply to such grants when made to Indian Tribes.

(3) A grant under paragraph (1) shall be made to cover not more than 75 per centum of the grant recipient's costs (as determined under regulations of the Administrator) in carrying out, during the one-year period beginning on the date the grant is made, an underground water source protection program.

(4) In each fiscal year the Administrator shall, in accordance with regulations, allot the sums appropriated for such year under paragraph (5) among the States on the basis of population, geographical area, and other relevant factors.

(5) For purposes of making grants under paragraph (1) there are authorized to be appropriated $5,000,000 for the fiscal year ending June 30, 1976, $7,500,000 for the fiscal year ending June 30, 1977, $10,000,000 for each of the fiscal years 1978 and 1979, $7,795,000 for the fiscal year ending September 30, 1980, $18,000,000 for the fiscal year ending September 30, 1981, and $21,000,000 for the fiscal year ending September 30, 1982. For the purpose of making grants under paragraph (1) there are authorized to be appropriated not more than the following amounts:

Fiscal year:

Amount

1987	$19,700,000
1988	19,700,000
1989	20,850,000
1990	20,850,000
1991	20,850,000
1992–2003	15,000,000.

(c) For purposes of this section:

(1) The term "public water system supervision program" means a program for the adoption and enforcement of drinking water regulations (with such variances and exemptions from such regulations under conditions and in a manner which is not less stringent than the conditions under, and the manner in, which variances and exemptions may be granted under sections 1415 and 1416) which are no less stringent than the national primary drinking water regulations under section 1412, and for keeping records and making reports required by section 1413(a)(3).

(2) The term "underground water source protection program" means a program for the adoption and enforcement of a program which meets the requirements of regulations under section 1421 and for keeping records and making reports required by section 1422(b)(1)(A)(ii). Such term includes, where applicable, a program which meets the requirements of section 1425.

(d) New York City Watershed Protection Program.—

(1) In General.—The Administrator is authorized to provide financial assistance to the State of New York for demonstration projects implemented as part of the watershed program for the protection and enhancement of the quality of source waters of the New York City water supply system, including projects that demonstrate, assess, or provide for comprehensive monitoring and surveillance and projects necessary to comply with the criteria for avoiding filtration contained in 40 CFR 141.71. Demonstration projects which shall be eligible for financial assistance shall be certified to the Administrator by the

State of New York as satisfying the purposes of this subsection. In certifying projects to the Administrator, the State of New York shall give priority to monitoring projects that have undergone peer review.

(2) Report.—Not later than 5 years after the date on which the Administrator first provides assistance pursuant to this paragraph, the Governor of the State of New York shall submit a report to the Administrator on the results of projects assisted.

(3) Matching Requirements.—Federal assistance provided under this subsection shall not exceed 50 percent of the total cost of the protection program being carried out for any particular watershed or ground water recharge area.

(4) Authorization.—There are authorized to be appropriated to the Administrator to carry out this subsection for each of fiscal years 2003 through 2010, $15,000,000 for the purpose of providing assistance to the State of New York to carry out paragraph (1).

Records and Inspections

Sec. 1445.

(a)

(1)

(A) Every person who is subject to any requirement of this title or who is a grantee, shall establish and maintain such records, make such reports, conduct such monitoring, and provide such information as the Administrator may reasonably require by regulation to assist the Administrator in establishing regulations under this title, in determining whether such person has acted or is acting in compliance with this title, in administering any program of financial assistance under this title, in evaluating the health risks of unregulated contaminants, or in advising the public of such risks. In requiring a public water system to monitor under this subsection, the Administrator may take into consideration the system size and the contaminants likely to be found in the system's drinking water.

(B) Every person who is subject to a national primary drinking water regulation under section 1412 shall provide such information as the Administrator may reasonably require, after consultation with the State in which such person is located if such State has primary enforcement responsibility for public water systems, on a case-by-case basis, to determine whether such person has acted or is acting in compliance with this title.

(C) Every person who is subject to a national primary drinking water regulation under section 1412 shall provide such information as the Administrator may reasonably require to assist the Administrator in establishing regulations under section 1412 of this title, after consultation with States and suppliers of water. The Administrator may not require under this subparagraph the installation of treatment equipment or process changes, the testing of treatment technology, or the analysis or processing of monitoring samples, except where the Administrator provides the funding for such activities. Before exercising this authority, the Administrator shall first seek to obtain the information by voluntary submission.

(D) The Administrator shall not later than 2 years after the date of enactment of this subparagraph, after consultation with public health experts, representatives of the general public, and officials of State and local governments, review the monitoring requirements for not fewer than 12 contaminants identified by the Administrator, and promulgate any necessary modifications.

(2) Monitoring Program for Unregulated Contaminants.—

(A) Establishment.—The Administrator shall promulgate regulations establishing the criteria for a monitoring program for unregulated contaminants. The regulations shall require monitoring of drinking water supplied by public water systems and shall vary the frequency and schedule for monitoring requirements for systems based on the number of persons served by the system, the source of supply, and the contaminants likely to be found, ensuring that only a representative sample of systems serving 10,000 persons or fewer are required to monitor.

(B) Monitoring Program for Certain UNREGULATED Contaminants.—

(i) Initial List.—Not later than 3 years after the date of enactment of the Safe Drinking Water Act Amendments of 1996 and every 5 years thereafter, the Administrator shall issue a list pursuant to subparagraph (A) of not more than 30 unregulated contaminants to be monitored by public water systems and to be included in the national drinking water occurrence data base maintained pursuant to subsection (g).

(ii) Governors' Petition.—The Administrator shall include among the list of contaminants for which monitoring is required under this paragraph each contaminant recommended in a petition signed by the Governor of each of 7 or more States, unless the Administrator determines that the action would prevent the listing of other contaminants of a higher public health concern.

(C) Monitoring Plan for Small and Medium Systems.—

(i) In General.—Based on the regulations promulgated by the Administrator, each State may develop a representative monitoring plan to assess the occurrence of unregulated contaminants in public water systems that serve a population of 10,000 or fewer in that State. The plan shall require monitoring for systems representative of different sizes, types, and geographic locations in the State.

(ii) Grants for Small System Costs.—From funds reserved under section 1452(o) or appropriated under subparagraph (H), the Administrator shall pay the reasonable cost of such testing and laboratory analysis as are necessary to carry out monitoring under the plan.

(D) Monitoring Results.—Each public water system that conducts monitoring of unregulated contaminants pursuant to this paragraph shall provide the results of the monitoring to the primary enforcement authority for the system.

(E) Notification.—Notification of the availability of the results of monitoring programs required under paragraph (2)(A) shall be given to the persons served by the system.

(F) Waiver of Monitoring Requirement.—The Administrator shall waive the requirement for monitoring for a contaminant under this paragraph in a State, if the State demonstrates that the criteria for listing the contaminant do not apply in that State.

(G) Analytical Methods.—The State may use screening methods approved by the Administrator under subsection (i) in lieu of monitoring for particular contaminants under this paragraph.

(H) Authorization of Appropriations.—There are authorized to be appropriated to carry out this paragraph $10,000,000 for each of the fiscal years [1997 through 2003] 2018 through 2022.

(b)

(1) Except as provided in paragraph (2), the Administrator, or representatives of the Administrator duly designated by him, upon presenting appropriate credentials and a written notice to any supplier of water or other person subject to (A) a national primary drinking water regulation prescribed under section 1412, (B) an applicable underground injection control program, or (C) any requirement to monitor an unregulated contaminant pursuant to subsection (a), or person in charge of any of the property of such supplier or other person referred to in clause (A), (B), or (C), is authorized to enter any establishment, facility, or other property of such supplier or other person in order to determine whether such supplier or other person has acted or is acting in compliance with this title, including for this purpose, inspection, at reasonable times, of records, files, papers, processes, controls, and facilities, or in order to test any feature

of a public water system, including its raw water source. The Administrator or the Comptroller General (or any representative designated by either) shall have access for the purpose of audit and examination to any records, reports, or information of a grantee which are required to be maintained under subsection (a) or which are pertinent to any financial assistance under this title.

(2) No entry may be made under the first sentence of paragraph (1) in an establishment, facility, or other property of a supplier of water or other person subject to a national primary drinking water regulation if the establishment, facility, or other property is located in a State which has primary enforcement responsibility for public water systems unless, before written notice of such entry is made, the Administrator (or his representative) notifies the State agency charged with responsibility for safe drinking water of the reasons for such entry. The Administrator shall, upon a showing by the State agency that such an entry will be detrimental to the administration of the State's program of primary enforcement responsibility, take such showing into consideration in determining whether to make such entry. No State agency which receives notice under this paragraph of an entry proposed to be made under paragraph (1) may use the information contained in the notice to inform the person whose property is proposed to be entered of the proposed entry; and if a State agency so uses such information, notice to the agency under this paragraph is not required until such time as the Administrator determines the agency has provided him satisfactory assurances that it will no longer so use information contained in a notice under this paragraph.

(c) Whoever fails or refuses to comply with any requirement of subsection (a) or to allow the Administrator, the Comptroller General, or representatives of either, to enter and conduct any audit or inspection authorized by subsection (b) shall be subject to a civil penalty of not to exceed $25,000.

(d)

(1) Subject to paragraph (2), upon a showing satisfactory to the Administrator by any person that any information required under this section from such person, if made public, would divulge trade secrets or secret processes of such person, the Administrator shall consider such information confidential in accordance with the purposes of section 1905 of title 18 of the United States Code. If the applicant fails to make a showing satisfactory to the Administrator, the Administrator shall give such applicant thirty days' notice before releasing the information to which the application relates (unless the public health or safety requires an earlier release of such information).

(2) Any information required under this section (A) may be disclosed to other officers, employees, or authorized representatives of the United States concerned with carrying out this title or to committees of the Congress, or

when relevant in any proceeding under this title, and (B) shall be disclosed to the extent it deals with the level of contaminants in drinking water. For purposes of this subsection the term "information required under this section" means any papers, books, documents, or information, or any particular part thereof, reported to or otherwise obtained by the Administrator under this section.

(e) For purposes of this section, (1) the term "grantee" means any person who applies for or receives financial assistance, by grant, contract, or loan guarantee under this title, and (2) the term "person" includes a Federal agency.

(f) Information Regarding Drinking Water Coolers.—The Administrator may utilize the authorities of this section for purposes of part F. Any person who manufactures, imports, sells, or distributes drinking water coolers in interstate commerce shall be treated as a supplier of water for purposes of applying the provisions of this section in the case of persons subject to part F.

(g) Occurrence Data Base.—

(1) In General.—Not later than 3 years after the date of enactment of the Safe Drinking Water Act Amendments of 1996, the Administrator shall assemble and maintain a national drinking water contaminant occurrence data base, using information on the occurrence of both regulated and unregulated contaminants in public water systems obtained under subsection (a)(1)(A) or subsection (a)(2) and reliable information from other public and private sources.

(2) Public Input.—In establishing the occurrence data base, the Administrator shall solicit recommendations from the Science Advisory Board, the States, and other interested parties concerning the development and maintenance of a national drinking water contaminant occurrence data base, including such issues as the structure and design of the data base, data input parameters and requirements, and the use and interpretation of data.

(3) Use.—The data shall be used by the Administrator in making determinations under section 1412(b)(1) with respect to the occurrence of a contaminant in drinking water at a level of public health concern.

(4) Public Recommendations.—The Administrator shall periodically solicit recommendations from the appropriate officials of the National Academy of Sciences and the States, and any person may submit recommendations to the Administrator, with respect to contaminants that should be included in the national drinking water contaminant occurrence data base, including recommendations with respect to additional unregulated contaminants that should be listed under subsection (a)(2). Any recommendation submitted under this clause shall be accompanied by reasonable documentation that—

(A) the contaminant occurs or is likely to occur in drinking water; and

(B) the contaminant poses a risk to public health.

(5) Public Availability.—The information from the data base shall be available to the public in readily accessible form.

(6) Regulated Contaminants.—With respect to each contaminant for which a national primary drinking water regulation has been established, the data base shall include information on the detection of the contaminant at a quantifiable level in public water systems (including detection of the contaminant at levels not constituting a violation of the maximum contaminant level for the contaminant).

(7) Unregulated Contaminants.—With respect to contaminants for which a national primary drinking water regulation has not been established, the data base shall include—

(A) monitoring information collected by public water systems that serve a population of more than 10,000, as required by the Administrator under subsection (a);

(B) monitoring information collected from a representative sampling of public water systems that serve a population of 10,000 or fewer; [and]

(C) if applicable, monitoring information collected by public water systems pursuant to subsection (j) that is not duplicative of monitoring information included in the data base under subparagraph (B) or (D); and

[(C)] (D) other reliable and appropriate monitoring information on the occurrence of the contaminants in public water systems that is available to the Administrator.

(h) Availability of Information on Small System Technologies.—For purposes of sections 1412(b)(4)(E) and 1415(e) (relating to small system variance program), the Administrator may request information on the characteristics of commercially available treatment systems and technologies, including the effectiveness and performance of the systems and technologies under various operating conditions. The Administrator may specify the form, content, and submission date of information to be submitted by manufacturers, States, and other interested persons for the purpose of considering the systems and technologies in the development of regulations or guidance under sections 1412(b)(4)(E) and 1415(e).

(i) Screening Methods.—The Administrator shall review new analytical methods to screen for regulated contaminants and may approve such methods as are more accurate or cost-effective than established reference methods for use in compliance monitoring.

(j) Monitoring by Certain Systems.—

(1) In General.—Notwithstanding subsection (a)(2)(A), the Administrator shall, subject to the availability of appropriations for such purpose—

(A) require public water systems serving between 3,300 and 10,000 persons to monitor for unregulated contaminants in accordance with this section; and

(B) ensure that only a representative sample of public water systems serving less than 3,300 persons are required to monitor.

(2) Effective Date.—Paragraph (1) shall take effect 3 years after the date of enactment of this subsection.

(3) Limitation.—Paragraph (1) shall take effect unless the Administrator determines that there is not sufficient laboratory capacity to accommodate the analysis necessary to carry out monitoring required under such paragraph.

(4) Authorization of Appropriations.—There are authorized to be appropriated $15,000,000 in each fiscal year for which monitoring is required to be carried out under this subsection for the Administrator to pay the reasonable cost of such testing and laboratory analysis as are necessary to carry out monitoring required under this subsection.

State Revolving Loan Funds

Sec. 1452.

(a) General Authority.—

(1) Grants to States to Establish State Loan Funds.—

(A) In General.—The Administrator shall offer to enter into agreements with eligible States to make capitalization grants, including letters of credit, to the States under this subsection to further the health protection objectives of this title, promote the efficient use of fund resources, and for other purposes as are specified in this title.

(B) Establishment of Fund.—To be eligible to receive a capitalization grant under this section, a State shall establish a drinking water treatment revolving loan fund (referred to in this section as a “State loan fund”) and comply with the other requirements of this section. Each grant to a State under this section shall be deposited in the State loan fund established by the State, except as otherwise provided in this section and in other provisions of this title. No funds authorized by other provisions of this title to be used for other purposes specified in this title shall be deposited in any State loan fund.

(C) Extended Period.—The grant to a State shall be available to the State for obligation during the fiscal year for which the funds are authorized and during the following fiscal year, except that grants made available from funds provided prior to fiscal year 1997 shall be available for obligation during each of the fiscal years 1997 and 1998.

(D) Allotment Formula.—Except as otherwise provided in this section, funds made available to carry out this section shall be allotted to States that have entered into an agreement pursuant to this section (other than the District of Columbia) in accordance with—
   (i) for each of fiscal years 1995 through 1997, a formula that is the same as the formula used to distribute public water system supervision grant funds under section 1443 in fiscal year 1995, except that the minimum proportionate share established in the formula shall be 1 percent of available funds and the formula shall be adjusted to include a minimum proportionate share for the State of Wyoming and the District of Columbia; and
   (ii) for fiscal year 1998 and each subsequent fiscal year, a formula that allocates to each State the proportional share of the State needs identified in the most recent survey conducted pursuant to subsection (h), except that the minimum proportionate share provided to each State shall be the same as the minimum proportionate share provided under clause (i).

(E) Reallotment.—The grants not obligated by the last day of the period for which the grants are available shall be reallotted according to the appropriate criteria set forth in subparagraph (D), except that the Administrator may reserve and allocate 10 percent of the remaining amount for financial assistance to Indian Tribes in addition to the amount allotted under subsection (i) and none of the funds reallotted by the Administrator shall be reallotted to any State that has not obligated all sums allotted to the State pursuant to this section during the period in which the sums were available for obligation.

(F) Nonprimacy States.—The State allotment for a State not exercising primary enforcement responsibility for public water systems shall not be deposited in any such fund but shall be allotted by the Administrator under this subparagraph. Pursuant to section 1443(a)(9)(A) such sums allotted under this subparagraph shall be reserved as needed by the Administrator to exercise primary enforcement responsibility under this title in such State and the remainder shall be reallotted to States exercising primary enforcement responsibility for public water systems for deposit in such funds. Whenever the Administrator makes a final determination pursuant to section 1413(b) that the requirements of section 1413(a) are no longer being met by a State, additional grants for such State under this title shall be immediately terminated by the Administrator. This subparagraph shall not apply to any State not exercising primary enforcement responsibility for public water systems as of the date of enactment of the Safe Drinking Water Act Amendments of 1996.

(G) Other Programs.—

(i) New System Capacity.—Beginning in fiscal year 1999, the Administrator shall withhold 20 percent of each capitalization grant made pursuant to this section to a State unless the State has met the requirements of section 1420(a) (relating to capacity development) and shall withhold 10 percent for fiscal year 2001, 15 percent for fiscal year 2002, and 20 percent for fiscal year 2003 if the State has not complied with the provisions of section 1420(c) (relating to capacity development strategies). Not more than a total of 20 percent of the capitalization grants made to a State in any fiscal year may be withheld under the preceding provisions of this clause. All funds withheld by the Administrator pursuant to this clause shall be reallotted by the Administrator on the basis of the same ratio as is applicable to funds allotted under subparagraph (D). None of the funds reallotted by the Administrator pursuant to this paragraph shall be allotted to a State unless the State has met the requirements of section 1420 (relating to capacity development).

(ii) Operator Certification.—The Administrator shall withhold 20 percent of each capitalization grant made pursuant to this section unless the State has met the requirements of 1419 (relating to operator certification). All funds withheld by the Administrator pursuant to this clause shall be reallotted by the Administrator on the basis of the same ratio as applicable to funds allotted under subparagraph (D). None of the funds reallotted by the Administrator pursuant to this paragraph shall be allotted to a State unless the State has met the requirements of section 1419 (relating to operator certification).

(2) Use of Funds.—

(A) In General.—Except as otherwise authorized by this title, amounts deposited in a State loan fund, including loan repayments and interest earned on such amounts, shall be used only for providing loans or loan guarantees, or as a source of reserve and security for leveraged loans, the proceeds of which are deposited in a State loan fund established under paragraph (1), or other financial assistance authorized under this section to community water systems and nonprofit noncommunity water systems, other than systems owned by Federal agencies.

(B) Limitation.—Financial assistance under this section may be used by a public water system only for expenditures [(including expenditures for planning, design, and associated preconstruction activities, including activities relating to the siting of the facility, but not] (including expenditures for planning, design, siting, and associated preconstruction activities, or for replacing or rehabilitating aging treatment, storage, or distribution facilities of public water systems, but not including monitoring, operation, and maintenance expenditures) of a type or category which the Administrator has determined, through guidance, will facilitate compliance with national primary drinking water regulations applicable to the system under section 1412 or otherwise significantly further the health protection objectives of this title.

(C) Sale of Bonds.—Funds may also be used by a public water system as a source of revenue (restricted solely to interest earnings of the applicable State loan fund) or security for payment of the principal and interest on revenue or general obligation bonds issued by the State to provide matching funds under subsection (e), if the proceeds of the sale of the bonds will be deposited in the State loan fund.

(D) Water Treatment Loans.—The funds under this section may also be used to provide loans to a system referred to in section 1401(4)(B) for the purpose of providing the treatment described in section 1401(4)(B)(i)(III).

(E) Acquisition of Real Property.—The funds under this section shall not be used for the acquisition of real property or interests therein, unless the acquisition is integral to a project authorized by this paragraph and the purchase is from a willing seller.

(F) Loan Assistance.—Of the amount credited to any State loan fund established under this section in any fiscal year, 15 percent shall be available solely for providing loan assistance to public water systems which regularly serve fewer than 10,000 persons to the extent such funds can be obligated for eligible projects of public water systems.

(3) Limitation.—

(A) In General.—Except as provided in subparagraph (B), no assistance under this section shall be provided to a public water system that—

(i) does not have the technical, managerial, and financial capability to ensure compliance with the requirements of this title; or

(ii) is in significant noncompliance with any requirement of a national primary drinking water regulation or variance.

(B) Restructuring.—A public water system described in subparagraph (A) may receive assistance under this section if—

(i) the use of the assistance will ensure compliance; and

(ii) if subparagraph (A)(i) applies to the system, the owner or operator of the system agrees to undertake feasible and appropriate changes in operations (including ownership, management, accounting, rates, maintenance, consolidation, alternative water supply, or other procedures) if the State determines that the measures are necessary to ensure that the system has the technical, managerial, and financial capability to comply with the requirements of this title over the long term.

(C) Review.—Prior to providing assistance under this section to a public water system that is in significant non-compliance with any requirement of a national primary drinking water regulation or variance, the State shall conduct a review to determine whether subparagraph (A)(i) applies to the system.

(4) American Iron and Steel Products.—

(A) In General.—During [fiscal year 2017] fiscal years 2018 through 2022, funds made available from a State loan fund established pursuant to this section may not be used for a project for the construction, alteration, or repair of a public water system unless all of the iron and steel products used in the project are produced in the United States.

(B) Definition Of Iron And Steel Products.—In this paragraph, the term "iron and steel products" means the following products made primarily of iron or steel:

(i) Lined or unlined pipes and fittings.
(ii) Manhole covers and other municipal castings.
(iii) Hydrants.
(iv) Tanks.
(v) Flanges.
(vi) Pipe clamps and restraints.
(vii) Valves.
(viii) Structural steel.
(ix) Reinforced precast concrete.
(x) Construction materials.

(C) Application.—Subparagraph (A) shall be waived in any case or category of cases in which the Administrator finds that—

(i) applying subparagraph (A) would be inconsistent with the public interest;
(ii) iron and steel products are not produced in the United States in sufficient and reasonably available quantities and of a satisfactory quality; or
(iii) inclusion of iron and steel products produced in the United States will increase the cost of the overall project by more than 25 percent.

(D) Waiver.—If the Administrator receives a request for a waiver under this paragraph, the Administrator shall make available to the public, on an informal basis, a copy of the request and information available to the Administrator concerning the request, and shall allow for informal public input on the request for at least 15 days prior to making a finding based on the request. The Administrator shall make the request and accompanying information available by electronic means, including on the official public Internet site of the Agency.

(E) International Agreements.—This paragraph shall be applied in a manner consistent with United States obligations under international agreements.

(F) Management and Oversight.—The Administrator may retain up to 0.25 percent of the funds appropriated for this section for management and oversight of the requirements of this paragraph.

(G) Effective Date.—This paragraph does not apply with respect to a project if a State agency approves the engineering plans and specifications for the project, in that agency's capacity to approve such plans and specifications prior to a project requesting bids, prior to the date of enactment of this paragraph.

(5) Evaluation.—During fiscal years 2018 through 2022, a State may provide financial assistance under this section to a public water system serving a population of more than 10,000 for an expenditure described in paragraph (2) only if the public water system—

(A) considers the cost and effectiveness of relevant processes, materials, techniques, and technologies for carrying out the project or activity that is the subject of the expenditure; and

(B) certifies to the State, in a form and manner determined by the State, that the public water system has made such consideration.

(6) Prevailing Wages.—The requirements of section 1450(e) shall apply to any construction project carried out in whole or in part with assistance made available by a drinking water treatment revolving loan fund.

(b) Intended Use Plans.—

(1) In General.—After providing for public review and comment, each State that has entered into a capitalization agreement pursuant to this section shall annually prepare a plan that identifies the intended uses of the amounts available to the State loan fund of the State.

(2) Contents.—An intended use plan shall include—

(A) a list of the projects to be assisted in the first fiscal year that begins after the date of the plan, including a description of the project, the expected terms of financial assistance, and the size of the community served;

(B) the criteria and methods established for the distribution of funds; and

(C) a description of the financial status of the State loan fund and the short-term and long-term goals of the State loan fund.

(3) Use of Funds.—

(A) In General.—An intended use plan shall provide, to the maximum extent practicable, that priority for the use of funds be given to projects that—

(i) address the most serious risk to human health;

(ii) are necessary to ensure compliance with the requirements of this title (including requirements for filtration); and

(iii) assist systems most in need on a per household basis according to State affordability criteria.

(B) List Of Projects.—Each State shall, after notice and opportunity for public comment, publish and periodically update a list of projects in the State that are eligible for assistance under this section, including the priority assigned to each project and, to the extent known, the expected funding schedule for each project.

(c) Fund Management.—Each State loan fund under this section shall be established, maintained, and credited with repayments and interest. The fund corpus shall be available in perpetuity for providing financial assistance under this section. To the extent amounts in the fund are not required for current obligation or expenditure, such amounts shall be invested in interest bearing obligations.

(d) Assistance for Disadvantaged Communities.—

(1) Loan Subsidy.—Notwithstanding any other provision of this section, in any case in which the State makes a loan pursuant to subsection (a)(2) to a disadvantaged community or to a community that the State expects to become a disadvantaged community as the result of a proposed project, the State may provide additional subsidization (including forgiveness of principal).

[(2) Total Amount of Subsidies.—For each fiscal year, the total amount of loan subsidies made by a State pursuant to paragraph (1) may not exceed 30 percent of the amount of the capitalization grant received by the State for the year.]

(2) Total Amount of Subsidies.—For each fiscal year, of the amount of the capitalization grant received by the State for the year, the total amount of loan subsidies made by a State pursuant to paragraph (1)—

(A) may not exceed 35 percent; and

(B) to the extent that there are sufficient applications for loans to communities described in paragraph (1), may not be less than 6 percent.

(3) Definition of Disadvantaged Community.—In this subsection, the term "disadvantaged community" means the service area of a public water system that meets affordability criteria established after public review and comment by the State in which the public water system is located. The Administrator may publish information to assist States in establishing affordability criteria.

(e) State Contribution.—Each agreement under subsection (a) shall require that the State deposit in the State loan fund from State moneys an amount equal to at least

20 percent of the total amount of the grant to be made to the State on or before the date on which the grant payment is made to the State, except that a State shall not be required to deposit such amount into the fund prior to the date on which each grant payment is made for fiscal years 1994, 1995, 1996, and 1997 if the State deposits the State contribution amount into the State loan fund prior to September 30, 1999.

(f) Types Of Assistance.—Except as otherwise limited by State law, the amounts deposited into a State loan fund under this section may be used only—

(1) to make loans, on the condition that—

(A) the interest rate for each loan is less than or equal to the market interest rate, including an interest free loan;

(B) principal and interest payments on each loan will commence not later than [1 year after completion of the project for which the loan was made, and each loan will be fully amortized not later than 20 years after the completion of the project, except that in the case of a disadvantaged community (as defined in subsection (d)(3)), a State may provide an extended term for a loan, if the extended term—]

[(i) terminates not later than the date that is 30 years after the date of project completion; and]

[(ii) does not exceed the expected design life of the project;] 18 months after completion of the project for which the loan was made;

(C) each loan will be fully amortized not later than 30 years after the completion of the project, except that in the case of a disadvantaged community (as defined in subsection (d)(3)) a State may provide an extended term for a loan, if the extended term—

(i) terminates not later than the date that is 40 years after the date of project completion; and

(ii) does not exceed the expected design life of the project;

[(C)] (D) the recipient of each loan will establish a dedicated source of revenue (or, in the case of a privately owned system, demonstrate that there is adequate security) for the repayment of the loan; and

[(D)] (E) the State loan fund will be credited with all payments of principal and interest on each loan;

(2) to buy or refinance the debt obligation of a municipality or an intermunicipal or interstate agency within the State at an interest rate that is less than or equal to the market interest rate in any case in which a debt obligation is incurred after July 1, 1993;

(3) to guarantee, or purchase insurance for, a local obligation (all of the proceeds of which finance a project eligible for assistance under this section) if the

guarantee or purchase would improve credit market access or reduce the interest rate applicable to the obligation;

(4) as a source of revenue or security for the payment of principal and interest on revenue or general obligation bonds issued by the State if the proceeds of the sale of the bonds will be deposited into the State loan fund; and

(5) to earn interest on the amounts deposited into the State loan fund.

(g) Administration of State Loan Funds.—

(1) Combined Financial Administration.—Notwithstanding subsection (c), a State may (as a convenience and to avoid unnecessary administrative costs) combine, in accordance with State law, the financial administration of a State loan fund established under this section with the financial administration of any other revolving fund established by the State if otherwise not prohibited by the law under which the State loan fund was established and if the Administrator determines that—

(A) the grants under this section, together with loan repayments and interest, will be separately accounted for and used solely for the purposes specified in subsection (a); and

(B) the authority to establish assistance priorities and carry out oversight and related activities (other than financial administration) with respect to assistance remains with the State agency having primary responsibility for administration of the State program under section 1413, after consultation with other appropriate State agencies (as determined by the State): Provided, That in non-primacy States eligible to receive assistance under this section, the Governor shall determine which State agency will have authority to establish priorities for financial assistance from the State loan fund.

(2) Cost of Administering Fund.—

(A) Authorization.—

(i) In General.—For each fiscal year, a State may use the amount described in clause (ii)—

(I) to cover the reasonable costs of administration of the programs under this section, including the recovery of reasonable costs expended to establish a State loan fund that are incurred after the date of enactment of this section; and

(II) to provide technical assistance to public water systems within the State.

(ii) Description of Amount.—The amount referred to in clause (i) is an amount equal to the sum of—

(I) the amount of any fees collected by the State for use in accordance with clause (i)(I), regardless of the source; and

(II) the greatest of—

(aa) $400,000;

(bb) 1/5 percent of the current valuation of the fund; and

(cc) an amount equal to 4 percent of all grant awards to the fund under this section for the fiscal year.

(B) Additional Use of Funds.—For fiscal year 1995 and each fiscal year thereafter, each State may use up to an additional 10 percent of the funds allotted to the State under this section—

(i) for public water system supervision programs under section 1443(a);

(ii) to administer or provide technical assistance through source water protection programs;

(iii) to develop and implement a capacity development strategy under section 1420(c); and

(iv) for an operator certification program for purposes of meeting the requirements of section 1419.

(C) Technical Assistance.—An additional 2 percent of the funds annually allotted to each State under this section may be used by the State to provide technical assistance to public water systems serving 10,000 or fewer persons in the State.

(D) Enforcement Actions.—Funds used under subparagraph (B)(ii) shall not be used for enforcement actions.

(3) Guidance and Regulations.—The Administrator shall publish guidance and promulgate regulations as may be necessary to carry out the provisions of this section, including—

(A) provisions to ensure that each State commits and expends funds allotted to the State under this section as efficiently as possible in accordance with this title and applicable State laws;

(B) guidance to prevent waste, fraud, and abuse; and

(C) guidance to avoid the use of funds made available under this section to finance the expansion of any public water system in anticipation of future population growth.

The guidance and regulations shall also ensure that the States, and public water systems receiving assistance under this section, use accounting, audit, and fiscal procedures that conform to generally accepted accounting standards.

(4) State Report.—Each State administering a loan fund and assistance program under this subsection shall publish and submit to the Administrator a report every 2 years on its activities under this section, including the findings of the most recent audit of the fund and the entire State allotment. The Administrator shall periodically audit all State loan funds established by, and all other

amounts allotted to, the States pursuant to this section in accordance with procedures established by the Comptroller General.

(h) Needs Survey.—[The Administrator]

(1) The Administrator shall conduct an assessment of water system capital improvement needs of all eligible public water systems in the United States and submit a report to the Congress containing the results of the assessment within 180 days after the date of enactment of the Safe Drinking Water Act Amendments of 1996 and every 4 years thereafter.

(2) Any assessment conducted under paragraph (1) after the date of enactment of the Drinking Water System Improvement Act of 2017 shall include an assessment of costs to replace all lead service lines (as defined in section 1459B(a)(4)) of all eligible public water systems in the United States, and such assessment shall describe separately the costs associated with replacing the portions of such lead service lines that are owned by an eligible public water system and the costs associated with replacing any remaining portions of such lead service lines, to the extent practicable.

(i) Indian Tribes.—

(1) In General.—1½ percent of the amounts appropriated annually to carry out this section may be used by the Administrator to make grants to Indian Tribes, Alaska Native villages, and, for the purpose of carrying out paragraph (5), intertribal consortia or tribal organizations, that have not otherwise received either grants from the Administrator under this section or assistance from State loan funds established under this section. Except as otherwise provided, the grants may only be used for expenditures by tribes and villages for public water system expenditures referred to in subsection (a)(2).

(2) Use of Funds.—Funds reserved pursuant to paragraph (1) shall be used to address the most significant threats to public health associated with public water systems that serve Indian Tribes, as determined by the Administrator in consultation with the Director of the Indian Health Service and Indian Tribes.

(3) Alaska Native Villages.—In the case of a grant for a project under this subsection in an Alaska Native village, the Administrator is also authorized to make grants to the State of Alaska for the benefit of Native villages. An amount not to exceed 4 percent of the grant amount may be used by the State of Alaska for project management.

(4) Needs Assessment.—The Administrator, in consultation with the Director of the Indian Health Service and Indian Tribes, shall, in accordance with a schedule that is consistent with the needs surveys conducted pursuant to subsection (h), prepare surveys and assess the needs of drinking water treatment facilities to serve Indian Tribes, including an evaluation of the public water systems that pose the most significant threats to public health.

(5) Training and Operator Certification.—

(A) In General.—The Administrator may use funds made available under this subsection and section 1442(e)(7) to make grants to intertribal consortia or tribal organizations for the purpose of providing operations and maintenance training and operator certification services to Indian Tribes to enable public water systems that serve Indian Tribes to achieve and maintain compliance with applicable national primary drinking water regulations.

(B) Eligible Tribal Organizations.—Intertribal consortia or tribal organizations eligible for a grant under subparagraph (A) are intertribal consortia or tribal organizations that—

(i) as determined by the Administrator, are the most qualified and experienced to provide training and technical assistance to Indian Tribes; and

(ii) the Indian Tribes find to be the most beneficial and effective.

(j) Other Areas.—Of the funds annually available under this section for grants to States, the Administrator shall make allotments in accordance with section 1443(a)(4) for the Virgin Islands, the Commonwealth of the Northern Mariana Islands, American Samoa, and Guam. The grants allotted as provided in this subsection may be provided by the Administrator to the governments of such areas, to public water systems in such areas, or to both, to be used for the public water system expenditures referred to in subsection (a)(2). The grants, and grants for the District of Columbia, shall not be deposited in State loan funds. The total allotment of grants under this section for all areas described in this subsection in any fiscal year shall not exceed 0.33 percent of the aggregate amount made available to carry out this section in that fiscal year.

(k) Other Authorized Activities.—

(1) In General.—Notwithstanding subsection (a)(2), a State may take each of the following actions:

(A) Provide assistance, only in the form of a loan, to one or more of the following:

(i) Any public water system described in subsection (a)(2) to acquire land or a conservation easement from a willing seller or grantor, if the purpose of the acquisition is to protect the source water of the system from contamination and to ensure compliance with national primary drinking water regulations.

(ii) Any community water system to implement local, voluntary source water protection measures to protect source water in areas delineated pursuant to section 1453, in order to facilitate compliance with national primary drinking water regulations applicable to the system under section 1412 or otherwise significantly further the health protection objectives of this title. Funds authorized under this clause may be used to fund only voluntary, incentive-based mechanisms. (iii) Any community water system to provide funding in accordance with section 1454(a)(1)(B)(i).

(B) Provide assistance, including technical and financial assistance, to any public water system as part of a capacity development strategy developed and implemented in accordance with section 1420(c).

(C) Make expenditures from the capitalization grant of the State [for fiscal years 1996 and 1997 to delineate and assess source water protection areas in accordance with section 1453] to delineate, assess, and update assessments for source water protection areas in accordance with section 1453, except that funds set aside for such expenditure shall be obligated within 4 fiscal years.

(D) Make expenditures from the fund for the establishment and implementation of wellhead protection programs under section 1428.

(2) Limitation.—For each fiscal year, the total amount of assistance provided and expenditures made by a State under this subsection may not exceed 15 percent of the amount of the capitalization grant received by the State for that year and may not exceed 10 percent of that amount for any one of the following activities:

(A) To acquire land or conservation easements pursuant to paragraph (1)(A)(i). to paragraph (1)(A)(i).

(B) To provide funding to implement voluntary, incentive-based source water quality protection measures pursuant to clauses (ii) and (iii) of paragraph (1)(A).

(C) To provide assistance through a capacity development strategy pursuant to paragraph (1)(B).

(D) To make expenditures to delineate or assess source water protection areas pursuant to paragraph (1)(C).

(E) To make expenditures to establish and implement wellhead protection programs pursuant to paragraph (1)(D).

(3) Statutory Construction.—Nothing in this section creates or conveys any new authority to a State, political subdivision of a State, or community water system for any new regulatory measure, or limits any authority of a State, political subdivision of a State or community water system.

(l) Savings.—The failure or inability of any public water system to receive funds under this section or any other loan or grant program, or any delay in obtaining the funds, shall not alter the obligation of the system to comply in a timely manner with all applicable drinking water standards and requirements of this title.

(m) Authorization of Appropriations.—[There are authorized to be appropriated to carry out the purposes of this section $599,000,000 for the fiscal year 1994 and $1,000,000,000 for each of the fiscal years 1995 through 2003.]

(1) There are authorized to be appropriated to carry out the purposes of this section—

(A) $1,200,000,000 for fiscal year 2018;

(B) $1,400,000,000 for fiscal year 2019;

(C) $1,600,000,000 for fiscal year 2020;

(D) $1,800,000,000 for fiscal year 2021; and

(E) $2,000,000,000 for fiscal year 2022. [To the extent amounts authorized to be]

(2) To the extent amounts authorized to be appropriated under this subsection in any fiscal year are not appropriated in that fiscal year, such amounts are authorized to be appropriated in a subsequent fiscal year [(prior to the fiscal year 2004)]. Such sums shall remain available until expended.

(n) Health Effects Studies.—From funds appropriated pursuant to this section for each fiscal year, the Administrator shall reserve $10,000,000 for health effects studies on drinking water contaminants authorized by the Safe Drinking Water Act Amendments of 1996. In allocating funds made available under this subsection, the Administrator shall give priority to studies concerning the health effects of cryptosporidium (as authorized by section 1458(c)), disinfection byproducts (as authorized by section 1458(c)), and arsenic (as authorized by section 1412(b)(12)(A)), and the implementation of a plan for studies of subpopulations at greater risk of adverse effects (as authorized by section 1458(a)).

(o) Monitoring for Unregulated Contaminants.—From funds appropriated pursuant to this section for each fiscal year beginning with fiscal year 1998, the Administrator shall reserve $2,000,000 to pay the costs of monitoring for unregulated contaminants under section 1445(a)(2)(C).

(p) Demonstration Project for State of Virginia.—Notwithstanding the other provisions of this section limiting the use of funds deposited in a State loan fund from any State allotment, the State of Virginia may, as a single demonstration and with the approval of the Virginia General Assembly and the Administrator, conduct a program to demonstrate alternative approaches to intergovernmental coordination to assist in the financing of new drinking water facilities in the following rural communities in southwestern Virginia where none exists on the date of enactment of the Safe Drinking Water Act Amendments of 1996 and where such communities are experiencing economic hardship: Lee County, Wise County, Scott County, Dickenson County, Russell County, Buchanan County, Tazewell County, and the city of Norton, Virginia. The funds allotted to that State and deposited in the State loan fund may be loaned to a regional endowment fund for the purpose set forth in this subsection under a plan to be approved by the Administrator. The plan may include an advisory group that includes representatives of such counties.

(q) Small System Technical Assistance.—The Administrator may reserve up to 2 percent of the total funds made available to carry out this section for each of fiscal years 2016 through 2021 to carry out the provisions of section 1442(e) (relating to technical assistance for small systems), except that the total amount of funds made available for such purpose in any fiscal year through appropriations (as authorized by section 1442(e)) and reservations made pursuant to this subsection shall not exceed the amount authorized by section 1442(e).

(r) Evaluation.—The Administrator shall conduct an evaluation of the effectiveness of the State loan funds through fiscal year 2001. The evaluation shall be submitted to the Congress at the same time as the President submits to the Congress, pursuant to section 1108 of title 31, United States Code, an appropriations request for fiscal year 2003 relating to the budget of the Environmental Protection Agency.

(s) Best Practices for State Loan Fund Administration.— The Administrator shall—

   (1) collect information from States on administration of State loan funds established pursuant to subsection (a)(1), including—

      (A) efforts to streamline the process for applying for assistance through such State loan funds;

      (B) programs in place to assist with the completion of applications for assistance through such State loan funds;

      (C) incentives provided to public water systems that partner with small public water systems to assist with the application process for assistance through such State loan funds;

      (D) practices to ensure that amounts in such State loan funds are used to provide loans, loan guarantees, or other authorized assistance in a timely fashion;

      (E) practices that support effective management of such State loan funds;

      (F) practices and tools to enhance financial management of such State loan funds; and

      (G) key financial measures for use in evaluating State loan fund operations, including—

         (i) measures of lending capacity, such as current assets and current liabilities or undisbursed loan assistance liability; and

         (ii) measures of growth or sustainability, such as return on net interest;

   (2) not later than 3 years after the date of enactment of the Drinking Water System Improvement Act of 2017, disseminate to the States best practices for administration of such State loan funds, based on the information collected pursuant to this subsection; and

(3) periodically update such best practices, as appropriate.

Source Water Petition Program

Sec. 1454.

(a) Petition Program.—

(1) In General.—

(A) Establishment.—A State may establish a program under which an owner or operator of a community water system in the State, or a municipal or local government or political subdivision of a State, may submit a source water quality protection partnership petition to the State requesting that the State assist in the local development of a voluntary, incentive-based partnership, among the owner, operator, or government and other persons likely to be affected by the recommendations of the partnership, to—

(i) reduce the presence in drinking water of contaminants that may be addressed by a petition by considering the origins of the contaminants, including to the maximum extent practicable the specific activities that affect the drinking water supply of a community;

(ii) obtain financial or technical assistance necessary to facilitate establishment of a partnership, or to develop and implement recommendations of a partnership for the protection of source water to assist in the provision of drinking water that complies with national primary drinking water regulations with respect to contaminants addressed by a petition; and

(iii) develop recommendations regarding voluntary and incentive-based strategies for the long-term protection of the source water of community water systems.

(B) Funding.—Each State may—

(i) use funds set aside pursuant to section 1452(k)(1)(A) (iii) by the State to carry out a program described in subparagraph (A), including assistance to voluntary local partnerships for the development and implementation of partnership recommendations for the protection of source water such as source water quality assessment, contingency plans, and demonstration projects for partners within a source water area delineated under section 1453(a); and

(ii) provide assistance in response to a petition submitted under this subsection using funds referred to in subsection (b)(2)(B).

(2) Objectives.—The objectives of a petition submitted under this subsection shall be to—

(A) facilitate the local development of voluntary, incentive-based partnerships among owners and operators of community water systems, governments, and other persons in source water areas; and

(B) obtain assistance from the State in identifying resources which are available to implement the recommendations of the partnerships to address the origins of drinking water contaminants that may be addressed by a petition (including to the maximum extent practicable the specific activities contributing to the presence of the contaminants) that affect the drinking water supply of a community.

(3) Contaminants Addressed by a Petition.—A petition submitted to a State under this subsection may address only those contaminants—

(A) that are pathogenic organisms for which a national primary drinking water regulation has been established or is required under section 1412; or

(B) for which a national primary drinking water regulation has been promulgated or proposed and that are detected by adequate monitoring methods in the source water at the intake structure or in any collection, treatment, storage, or distribution facilities by the community water systems at levels—

(i) above the maximum contaminant level; or

(ii) that are not reliably and consistently below the maximum contaminant level.

(4) Contents.—A petition submitted under this subsection shall, at a minimum—

(A) include a delineation of the source water area in the State that is the subject of the petition;

(B) identify, to the maximum extent practicable, the origins of the drinking water contaminants that may be addressed by a petition (including to the maximum extent practicable the specific activities contributing to the presence of the contaminants) in the source water area delineated under section 1453;

(C) identify any deficiencies in information that will impair the development of recommendations by the voluntary local partnership to address drinking water contaminants that may be addressed by a petition;

(D) specify the efforts made to establish the voluntary local partnership and obtain the participation of—

(i) the municipal or local government or other political subdivision of the State with jurisdiction over the source water area delineated under section 1453; and

(ii) each person in the source water area delineated under section 1453—

(I) who is likely to be affected by recommendations of the voluntary local partnership; and

(II) whose participation is essential to the success of the partnership;

(E) outline how the voluntary local partnership has or will, during development and implementation of recommendations of the voluntary local partnership, identify, recognize and take into account any voluntary or other activities already being undertaken by persons in the source water area delineated under section 1453 under Federal or State law to reduce the likelihood that contaminants will occur in drinking water at levels of public health concern; and

(F) specify the technical, financial, or other assistance that the voluntary local partnership requests of the State to develop the partnership or to implement recommendations of the partnership.

(b) Approval or Disapproval of Petitions.—

(1) In General.—After providing notice and an opportunity for public comment on a petition submitted under subsection (a), the State shall approve or disapprove the petition, in whole or in part, not later than 120 days after the date of submission of the petition.

(2) Approval.—The State may approve a petition if the petition meets the requirements established under subsection (a). The notice of approval shall, at a minimum, include for informational purposes—

(A) an identification of technical, financial, or other assistance that the State will provide to assist in addressing the drinking water contaminants that may be addressed by a petition based on—

(i) the relative priority of the public health concern identified in the petition with respect to the other water quality needs identified by the State;

(ii) any necessary coordination that the State will perform of the program established under this section with programs implemented or planned by other States under this section; and

(iii) funds available (including funds available from a State revolving loan fund established under title VI of the Federal Water Pollution Control Act (33 U.S.C. 1381 et seq.)) or section 1452;

(B) a description of technical or financial assistance pursuant to Federal and State programs that is available to assist in implementing recommendations of the partnership in the petition, including—

(i) any program established under the Federal Water Pollution Control Act (33 U.S.C. 1251 et seq.); (ii) the program established under section 6217 of the Coastal Zone Act Reauthorization Amendments of 1990 (16 U.S.C. 1455b);

(iii) the agricultural water quality protection program established under chapter 2 of subtitle D of title XII of the Food Security Act of 1985 (16 U.S.C. 3838 et seq.);

(iv) the sole source aquifer protection program established under section 1427;

(v) the community wellhead protection program established under section 1428;

(vi) any pesticide or ground water management plan;

(vii) any voluntary agricultural resource management plan or voluntary whole farm or whole ranch management plan developed and implemented under a process established by the Secretary of Agriculture; and

(viii) any abandoned well closure program; and

(C) a description of activities that will be undertaken to coordinate Federal and State programs to respond to the petition.

(3) Disapproval.—If the State disapproves a petition submitted under subsection (a), the State shall notify the entity submitting the petition in writing of the reasons for disapproval. A petition may be resubmitted at any time if—

(A) new information becomes available;

(B) conditions affecting the source water that is the subject of the petition change; or

(C) modifications are made in the type of assistance being requested.

(c) Grants to Support State Programs.—

(1) In General.—The Administrator may make a grant to each State that establishes a program under this section that is approved under paragraph (2). The amount of each grant shall not exceed 50 percent of the cost of administering the program for the year in which the grant is available.

(2) Approval.—In order to receive grant assistance under this subsection, a State shall submit to the Administrator for approval a plan for a source water quality protection partnership program that is consistent with the guidance published under subsection (d). The Administrator shall approve the plan if the plan is consistent with the guidance published under subsection (d).

(d) Guidance.—

(1) In General.—Not later than 1 year after the date of enactment of this section, the Administrator, in consultation with the States, shall publish guidance to assist—

(A) States in the development of a source water quality protection partnership program; and

(B) municipal or local governments or political subdivisions of a State and community water systems in the development of source water quality protection partnerships and in the assessment of source water quality.

(2) Contents of the Guidance.—The guidance shall, at a minimum—

(A) recommend procedures for the approval or disapproval by a State of a petition submitted under subsection (a);

(B) recommend procedures for the submission of petitions developed under subsection (a);

(C) recommend criteria for the assessment of source water areas within a State; and

(D) describe technical or financial assistance pursuant to Federal and State programs that is available to address the contamination of sources of drinking water and to develop and respond to petitions submitted under subsection (a).

(e) Authorization of Appropriations.—There are authorized to be appropriated to carry out this section $5,000,000 for each of the fiscal years ø1997 through 2003¿ 2018 through 2022. Each State with a plan for a program approved under subsection (b) shall receive an equitable portion of the funds available for any fiscal year.

(f) Statutory Construction.—Nothing in this section—

(1)

(A) creates or conveys new authority to a State, political subdivision of a State, or community water system for any new regulatory measure; or

(B) limits any authority of a State, political subdivision, or community water system; or

(2) precludes a community water system, municipal or local government, or political subdivision of a government from locally developing and carrying out a voluntary, incentive-based, source water quality protection partnership to address the origins of drinking water contaminants of public health concern.

Sec. 1459c. Review of Technologies.

(a) Review.—The Administrator, after consultation with appropriate departments and agencies of the Federal Government and with State and local governments, shall review (or enter into contracts or cooperative agreements to provide for a review of) existing and potential methods, means, equipment, and technologies (including review of cost, availability, and efficacy of such methods, means, equipment, and technologies) that—

(1) ensure the physical integrity of community water systems;

(2) prevent, detect, and respond to any contaminant for which a national primary drinking water regulation has been promulgated in community water systems and source water for community water systems;

(3) allow for use of alternate drinking water supplies from nontraditional sources; and

(4) facilitate source water assessment and protection.

(b) Inclusions.—The review under subsection (a) shall include review of methods, means, equipment, and technologies—

(1) that are used for corrosion protection, metering, leak detection, or protection against water loss;

(2) that are intelligent systems, including hardware, software, or other technology, used to assist in protection and detection described in paragraph (1);

(3) that are point-of-use devices or point-of-entry devices;

(4) that are physical or electronic systems that monitor, or assist in monitoring, contaminants in drinking water in real-time; and

(5) that allow for the use of nontraditional sources for drinking water, including physical separation and chemical and biological transformation technologies.

(c) Availability.—The Administrator shall make the results of the review under subsection (a) available to the public.

(d) Authorization of Appropriations.—There are authorized to be appropriated to the Administrator to carry out this section $10,000,000 for fiscal year 2018, which shall remain available until expended.

Part F—Additional Requirements to Regulate the Safety of Drinking Water

Definitions

Sec. 1461. As used in this part—

(1) Drinking Water Cooler.—The term "drinking water cooler" means any mechanical device affixed to drinking water supply plumbing which actively cools water for human consumption.

(2) Lead Free.—The term "lead free" means, with respect to a drinking water cooler, that each part or component of the cooler which may come in contact with drinking water contains not more than 8 percent lead, except that no drinking water cooler which contains any solder, flux, or storage tank interior surface which may come in contact with drinking water shall be considered lead free if the solder, flux, or storage tank interior surface contains more than 0.2 percent lead. The Administrator may establish more stringent requirements for treating any part or component of a drinking water cooler as lead free for purposes of this part

whenever he determines that any such part may constitute an important source of lead in drinking water.

(3) Local Educational Agency.—The term "local educational agency" means—
  (A) any local educational agency as defined in section 8101 of the Elementary and Secondary Education Act of 1965,
  (B) the owner of any private, nonprofit elementary or secondary school building, and
  (C) the governing authority of any school operating under the defense dependent's education system provided for under the Defense Dependent's Education Act of 1978 (20 U.S.C. 921 and following).

(4) Repair.—The term "repair" means, with respect to a drinking water cooler, to take such corrective action as is necessary to ensure that water cooler is lead free.

(5) Replacement.—The term "replacement", when used with respect to a drinking water cooler or drinking water fountain, means the permanent removal of the water cooler or drinking water fountain and the installation of a lead free water cooler or drinking water fountain.

(6) School.—The term "school" means any elementary school or secondary school as defined in section 8101 of the Elementary and Secondary Education Act of 1965 and any kindergarten or day care facility.

(7) Lead-lined Tank.—The term "lead-lined tank" means a water reservoir container in a drinking water cooler which container is constructed of lead or which has an interior surface which is not leadfree.

Sec. 1465. Drinking Water Fountain Replacement for Schools.

(a) ESTABLISHMENT.—Not later than 1 year after the date of enactment of this section, the Administrator shall establish a grant program to provide assistance to local educational agencies for the replacement of drinking water fountains manufactured prior to 1988.

(b) USE OF FUNDS.—Funds awarded under the grant program—
  (1) shall be used to pay the costs of replacement of drinking water fountains in schools; and
  (2) may be used to pay the costs of monitoring and reporting of lead levels in the drinking water of schools of a local educational agency receiving such funds, as determined appropriate by the Administrator.

(c) Priority.—In awarding funds under the grant program, the Administrator shall give priority to local educational agencies based on economic need.

(d) Authorization of Appropriations.—There are authorized to be appropriated to carry out this section not more than $5,000,000 for each of fiscal years 2018 through 2022.

Emergency Planning and Community Right-to- Know Act of 1986

Title III—Emergency Planning and Community Right-to-Know

Subtitle A—Emergency Planning and Notification

Sec. 304. Emergency Notification.

(a) Types of Releases.—

(1) 302 302(A) substance which requires cercla notice.—If a release of an extremely hazardous substance referred to in section 302(a) occurs from a facility at which a hazardous chemical is produced, used, or stored, and such release requires a notification under section 103(a) of the Comprehensive Environmental Response, compensation, and Liability Act of 1980 (hereafter in this section referred to as "CERCLA") (42 U.S.C. 9601 et seq.), the owner or operator of the facility shall immediately provide notice as described in subsection (b).

(2) Other 302(A) substance.—If a release of an extremely hazardous substance referred to in section 302(a) occurs from a facility at which a hazardous chemical is produced, used, or stored, and such release is not subject to the notification requirements under section 103(a) of CERCLA, the owner or operator of the facility shall immediately provide notice as described in subsection (b), but only if the release—

(A) is not a federally permitted release as defined in section 101(10) of CERCLA,

(B) is in an amount in excess of a quantity which the Administrator has determined (by regulation) requires notice, and (C) occurs in a manner which would require notification under section 103(a) of CERCLA.

Unless and until superseded by regulations establishing a quantity for an extremely hazardous substance described in this paragraph, a quantity of 1 pound shall be deemed that quantity the release of which requires notice as described in subsection (b).

(3) Non-302 Non-302(A) substance which requires cercla notice.—If a release of a substance which is not on the list referred to in section 302(a) occurs at a facility at which a hazardous chemical is produced, used, or stored, and such

release requires notification under section 103(a) of CERCLA, the owner or operator shall provide notice as follows:

(A) If the substance is one for which a reportable quantity has been established under section 102(a) of CERCLA, the owner or operator shall provide notice as described in subsection (b).

(B) If the substance is one for which a reportable quantity has not been established under section 102(a) of CERCLA—

(i) Until April 30, 1988, the owner or operator shall provide, for releases of one pound or more of the substance, the same notice to the community emergency coordinator for the local emergency planning committee, at the same time and in the same form, as notice is provided to the National Response Center under section 103(a) of CERCLA.

(ii) On and after April 30,1988, the owner or operator shall provide, for releases of one pound or more of the substance, the notice as described in subsection (b).

(4) Exempted Leases.—This section does not apply to any release which results in exposure to persons solely within the site or sites on which a facility is located.

(b) Notification.—

(1) Recipient of Notice.—Notice required under subsection (a) shall be given immediately after the release by the owner or operator of a facility (by such means as telephone, radio, or in person) to the community emergency coordinator for the local emergency planning committees, if established pursuant to section 301(c), for any area likely to be affected by the release and to the [State emergency planning commission] State emergency response commission of any State likely to be affected by the release. With respect to transportation of a substance subject to the requirements of this section, or storage incident to such transportation, the notice requirements of this section with respect to a release shall be satisfied by dialing 911 or, in the absence of a 911 emergency telephone number, calling the operator.

(2) Contents.—Notice required under subsection (a) shall include each of the following (to the extent known at the time of the notice and so long as no delay in responding to the emergency results):

(A) The chemical name or identity of any substance involved in the release.

(B) An indication of whether the substance is on the list referred to in section 302(a).

(C) An estimate of the quantity of any such substance that was released into the environment.

(D) The time and duration of the release.

(E) The medium or media into which the release occurred.

(F) Any known or anticipated acute or chronic health risks associated with the emergency and, where appropriate, advice regarding medical attention necessary for exposed individuals.

(G) Proper precautions to take as a result of the release, including evacuation (unless such information is readily available to the community emergency coordinator pursuant to the emergency plan).

(H) The name and telephone number of the person or persons to be contacted for further information.

(c) Followup Emergency Notice.—AS soon as practicable after a release which requires notice under subsection (a), such owner or operator shall provide a written followup emergency notice (or notices, as more information becomes available) setting forth and updating the information required under subsection (b), and including additional information with respect to—

(1) actions taken to respond to and contain the release,

(2) any known or anticipated acute or chronic health risks associated with the release, and

(3) where appropriate, advice regarding medical attention necessary for exposed individuals.

(d) Transportation Exemption not Applicable.—The exemption provided in section 327 (relating to transportation) does not apply to this section.

(e) Addressing Source Water Used for Drinking Water.—

(1) Applicable State Agency Notification.—A State emergency response commission shall—

(A) promptly notify the applicable State agency of any release that requires notice under subsection (a);

(B) provide to the applicable State agency the information identified in subsection (b)(2); and

(C) provide to the applicable State agency a written followup emergency notice in accordance with subsection (c).

(2) Community Water System Notification.—

(A) In General.—An applicable State agency receiving notice of a release under paragraph (1) shall—

(i) promptly forward such notice to any community water system the source waters of which are affected by the release;

(ii) forward to the community water system the information provided under paragraph (1)(B); and

(iii) forward to the community water system the written followup emergency notice provided under paragraph (1)(C).

(B) Direct Notification.—In the case of a State that does not have an applicable State agency, the State emergency response commission shall provide the notices and information described in paragraph (1) directly to any community water system the source waters of which are affected by a release that requires notice under subsection (a).

(3) Definitions.—In this subsection:

(A) Community Water System.—The term “community water system” has the meaning given such term in section 1401(15) of the Safe Drinking Water Act.

(B) Applicable State Agency.—The term “applicable State agency” means the State agency that has primary responsibility to enforce the requirements of the Safe Drinking Water Act in the State.

Subtitle B—Reporting Requirements

Sec. 312. Emergency and Hazardous Chemical Inventory Forms.

(a) Basic Requirement.—

(1) The owner or operator of any facility which is required to prepare or have available a material safety data sheet for a hazardous chemical under the Occupational Safety and Health Act of 1970 and regulations promulgated under that Act shall prepare and submit an emergency and hazardous chemical inventory form (hereafter in this title referred to as an “inventory form”) to each of the following:

(A) The appropriate local emergency planning committee.

(B) The State emergency response commission.

(C) The fire department with jurisdiction over the facility.

(2) The inventory form containing tier I information (as described in subsection (d)(l)) shall be submitted on or before March 1, 1988, and annually thereafter on March 1, and shall contain data with respect to the preceding calendar year. The preceding sentence does not apply if an owner or operator provides, by the same deadline and with respect to the same calendar year, tier 11 information (as described in subsection (d)(2)) to the recipients described in paragraph (1).

(3) An owner or operator may meet the requirements of this section with respect to a hazardous chemical which is a mixture by doing one of the following:

(A) Providing information on the inventory form on each element or compound in the mixture which is a hazardous chemical. If more than one mixture has the same element or compound, only one listing on the inventory form for the element or compound at the facility is necessary.

(B) Providing information on the inventory form on the mixture itself.

(b) Thresholds.—The Administrator may establish threshold quantities for hazardous chemicals covered by this section below which no facility shall be subject to the provisions of this section. The threshold quantities may, in the Administrator's discretion, be based on classes of chemicals or categories of facilities.

(c) Hazardous Chemicals Covered.—A hazardous chemical subject to the requirements of this section is any hazardous chem-planning commission¿ State emergency response commission, a under section 311.

(d) Contents of Form.—

(1) Tier i Information.—

(A) Aggregate Information by Category.—An inventory form shall provide the information described in subparagraph (B) in aggregate terms for hazardous chemicals in categories of health and physical hazards as set forth under the Occupational Safety and Health Act of 1970 and regulations promulgated under that Act.

(B) Required Information.—The information referred to in subparagraph (A) is the following:

(i) An estimate (in ranges) of the maximum amount of hazardous chemicals in each category present at the facility at any time during the preceding calendar year.

(ii) An estimate (in ranges) of the average daily amount of hazardous chemicals in each category present at the facility during the preceding calendar year.

(iii) The general location of hazardous chemicals in each category.

(C) Modifications.—For purposes of reporting information under this paragraph, the Administrator may—

(i) modify the categories of health and physical hazards as set forth under the Occupational Safety and Health Act of 1970 and regulations promulgated under that Act by requiring information to be reported in terms of groups of hazardous chemicals which present similar hazards in an emergency, or

(ii) require reporting on individual hazardous chemicals of special concern to emergency response personnel.

(2) Tier ii Information.—An inventory form shall provide the following additional information for each hazardous chemical present at the facility, but only upon request and in accordance with subsection (e):

(A) The chemical name or the common name of the chemical as provided on the material safety data sheet.

(B) An estimate (in ranges) of the maximum amount of the hazardous chemical present at the facility at any time during the preceding calendar year.

(C) An estimate (in ranges) of the average daily amount of the hazardous chemical present at the facility during the preceding calendar year.

(D) A brief description of the manner of storage of the hazardous chemical.

(E) The location at the facility of the hazardous chemical.

(F) An indication of whether the owner elects to with-hold location information of a specific hazardous chemical from disclosure to the public under section 324.

(e) Availability of Tier ii Information.—

(1) Availability to State Commissions, Local Committees, and Fire Departments.—Upon request by a [State emergency planning commission] State emergency response commission, a local emergency planning committee, or a fire department with jurisdiction over the facility, the owner or operator of a facility shall provide tier II information, as described in subsection (d), to the person making the request. Any such request shall be with respect to a specific facility.

(2) Availability to Other State and Local Officials.—A State or local official acting in his or her official capacity may have access to tier II information by submitting a request to the State emergency response commission or the local emergency planning committee. Upon receipt of a request for tier II information, the State commission or local committee shall, pursuant to paragraph (1), request the facility owner or operator for the tier II information and make available such information to the official.

(3) Availability to Public.—

(A) In General.—Any person may request a State emergency response commission or local emergency planning committee for tier II information relating to the preceding calendar year with respect to a facility. Any such request shall be in writing and shall be with respect to a specific facility.

(B) Automatic Provision of Information to Public.— Any tier II information which a State emergency response commission or local emergency planning committee has in its possession shall be made available to a person making a request under this paragraph in accordance with section 324. If the State emergency response commission or local emergency planning committee does not have the tier II information in its possession, upon a request for tier II information the State emergency response commission or local emergency planning committee shall, pursuant to paragraph (1), request the facility owner or operator for tier II information with respect to a hazardous chemical which a facility has stored in an amount in excess of 10,000 pounds present at the facility at any time during the preceding calendar year and make such information available in accordance with section 324 to the person making the request.

(C) Discretionary Provision of Information to Public.— In the case of tier II information which is not in the possession of a State emergency response commission or local emergency planning committee and which is with respect to a hazardous chemical which a facility has stored in an amount less than 10,000 pounds present at the facility at any time during the preceding calendar year, a request from a person must include the general need for the information. The State emergency response commission or local emergency planning committee may, pursuant to paragraph (1), request the facility owner or operator for the tier II information on behalf of the person making the request. Upon receipt of any information requested on behalf of such person, the State emergency response commission or local emergency planning committee shall make the information available in accordance with section 324 to the person.

(D) Response in 45 Days.—A State emergency response commission or local emergency planning committee shall respond to a request for tier II information under this paragraph no later than 45 days after the date of receipt of the request.

(4) Availability to Community Water Systems.—

(A) In General.—An affected community water system may have access to tier II information by submitting a request to the State emergency response commission or the local emergency planning committee. Upon receipt of a request for tier II information, the State commission or local committee shall, pursuant to paragraph (1), request the facility owner or operator for the tier II information and make available such information to the affected community water system.

(B) Definition.—In this paragraph, the term "affected community water system" means a community water system (as defined in section 1401(15) of the Safe Drinking Water Act) that receives supplies of drinking water from a source water area, delineated under section 1453 of the Safe Drinking Water Act, in which a facility that is required to prepare and submit an inventory form under subsection (a)(1) is located.

(f) Fire Department Access.—Upon request to an owner or operator of a facility which files an inventory form under this section by the fire department with jurisdiction over the facility, the owner or operator of the facility shall allow the fire department to conduct an on-site inspection of the facility and shall provide to the fire department specific location information on hazardous chemicals at the facility.

(g) Format of Forms.—The Administrator shall publish a uniform format for inventory forms within three months after the date of the enactment of this title. If the Administrator does not publish such forms, owners and operators of facilities subject to the requirements of this section shall provide the information required under this section by letter.

ADDITIONAL VIEWS

Our nation's public drinking water systems serve over 300 million people, but aging and failing infrastructure threatens reliable access to safe drinking water. Earlier this year, the American Society of Civil Engineers published its periodic infrastructure report card, and rated our drinking water infrastructure a "D" grade.[32] Most of the pipes in this country are between 75 and 110 years old—at or beyond the expected limits of their useful life. Because of this, an estimated 240,000 water main breaks occur every year, wasting money, disrupting service, and compromising water quality.[33]

Lead is also a significant and growing threat from our aging infrastructure. Lead is present in our drinking water distribution systems in service lines, solder, and fixtures. As that infrastructure ages and corrodes, more lead can leach into drinking water.

To maintain safe drinking water delivery, public water systems will need to make significant investments to repair or replace infrastructure and equipment. EPA's most recent needs assessment for drinking water infrastructure estimated that $384 billion will be necessary for infrastructure repairs by 2030.[34] This amount grew significantly since the agency's last assessment, demonstrating that investment has not kept pace with need.[35]

Delaying these investments will increase costs because repairing damaged pipes is more expensive than replacing them before breakage.[36] Old pipes will continue to break resulting in massive quantities of lost treated water, and prompting inefficient emergency repair expenditures. These costs are then passed onto the consumer in higher utility bills and increased service disruptions. We support the Drinking Water System Improvement Act, but believe that higher funding levels will ultimately be needed to address our drinking water infrastructure needs.

Over the course of the Committee's consideration of this bill, provisions were added to address several serious drinking water challenges, including resiliency to extreme weather and intentional attacks, improved consumer notification requirements, and

---

[32] American Society of Civil Engineers, *2017 Report Card for America's Infrastructure,* (online at www.infrastructurereportcard.org).

[33] *Id.*

[34] U.S. Environmental Protection Agency, *Drinking Water Infrastructure Needs Survey and Assessment, Fifth Report to Congress* (April 2013) (EPA–816–R–13–006) (online at water.epa.gov/grants_funding/dwsrf/index.cfm).

[35] *Id.*

[36] *Id.*

improved monitoring. We strongly support these provisions, but note that some serious drinking water challenges are still not addressed by the bill. In particular, the onerous and unworkable standard setting process in place since the 1996 Safe Drinking Water Act Amendments is not changed by this bill. That ineffectual process has prevented EPA from revising and setting needed drinking water standards, including standards for lead, perchlorate, perfluorinated chemicals, and algal toxins.

Additionally, the impacts of Hurricanes Irma and Maria in Puerto Rico and the U.S. Virgin Islands show that the federal government must take a more active role in addressing drinking water needs throughout all of our nation. As we write, more than one month after Hurricane Maria hit Puerto Rico, roughly a quarter of the residents of Puerto Rico have no access to drinking water. Providing access to safe drinking water is a fundamental function of our government, and we must do more to meet our responsibility in U.S. territories.

We hope to build on the Drinking Water System Improvement Act in the coming months and years to address these remaining challenges and remaining funding needs.

Frank Pallone, Jr.,
Ranking Member, Committee on Energy and Commerce.

Paul D. Tonko,
Ranking Member,
Subcommittee on Environment.

INDEX

A

B

C

D

E

F

G

H

I

J

L

M

N

O

P

Q

R

S

T

U

V

W

Related Nova Publications

American Recovery and Reinvestment Act: Economic Impact Five Years Later

Editor: Elise Taylor

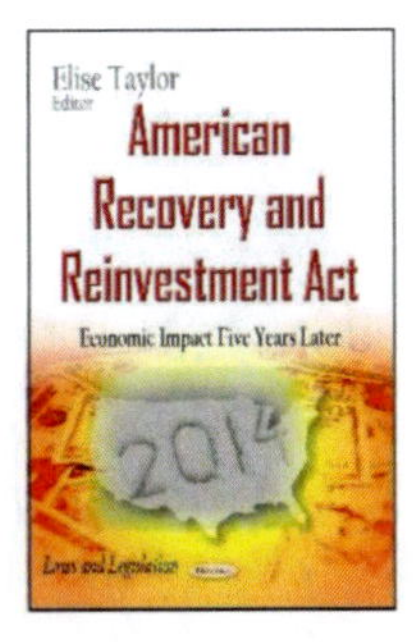

Series: Laws and Legislation

Book Description: In February 2009, in response to significant weakness in the economy, lawmakers enacted the American Recovery and Reinvestment Act (ARRA). This book discusses the economic impact of the ARRA five years after its enactment.

Softcover ISBN: 978-1-63321-395-1
Retail Price: $69

The U.S. Minimum Wage: Issues and Potential Effects of an Increase

Editor: Joel Biondi

Series: Economic Issues, Problems and Perspectives

Book Description: This book discusses the federal minimum wage; inflation and the effect it has on minimum wage; the effects on employment and family income of an increase of minimum-wage; and the Fair Labor Standards Act (FLSA), which is the federal legislation that establishes the minimum hourly wage that must be paid to all covered workers.

Softcover ISBN: 978-1-63117-689-0
Retail Price: $58